A Compendium of Solid State Theory

Ladislaus Alexander Bányai

A Compendium of Solid State Theory

 Springer

Ladislaus Alexander Bányai
Oberursel
Germany

ISBN 978-3-030-08744-9   ISBN 978-3-319-78613-1   (eBook)
https://doi.org/10.1007/978-3-319-78613-1

Printed on acid-free paper

This Springer imprint is published by the registered company Springer International Publishing AG
part of Springer Nature
The registered company address is: Gewerbestrasse 11, 6330 Cham, Switzerland

Preface

This compendium emerged from my lecture notes at the Physics Department of the Johann Wolfgang Goethe University in Frankfurt am Main till 2004 and does not include recent progresses in the field. It is less than a textbook, but rather more than a German "Skript". It does not include a bibliography or comparison with experiments. Mathematical proofs are often only sketched. Nevertheless, it may be useful to graduate students as a concise presentation of the basics of solid-state theory.

Oberursel, Germany   Ladislaus Alexander Bányai

Contents

Chapter 1
Introduction

A short presentation is given about how we conceive theoretically the solid state of matter, namely about its constituents and their interactions. The starting point is of course an oversimplified one, which afterwards one tries to improve and even redefine it. In the opposite sense, one tries later to simplify it in order to allow for theoretical calculations. In spite of all these problems, one gets predictions, that often may be successfully confronted with experiments. The different Chapters of this Compendium are however intended only to introduce the reader into the basic concepts of solid state theory, without the usual detours about specific materials and without comparison with experiments.

Under Solid State we understand a stable macroscopic cluster of atoms. The stability of this system relies on the interaction between its constituents. We know today, that molecules and atoms are built up of protons, neutrons and electrons. According to the modern fundamental concepts of matter at their turn these particles are made up however of some other more elementary ones we do not need to list here. The simplest starting point of Solid State theory and still the only useful one is, that we have a quantum mechanical system of ions and valence electrons with Coulomb interactions between these particles. This non-relativistic picture of a system of charged particles however is valid only up to effects of order $\frac{1}{c^2}$ (*c*-being the light velocity in vacuum). Actually, the charged particles are themselves sources of an electromagnetic field, which on its turn is quantized (photons). Fields and particles have to be treated consequently together as a single system. Nevertheless, most of the properties of solid state are successfully described by a non-relativistic quantum mechanical Hamiltonian

$$H = H_e + H_i + H_{ee} + H_{ii} + H_{ei} .$$

© Springer International Publishing AG, part of Springer Nature 2018

L. A. Bányai, *A Compendium of Solid State Theory*,

https://doi.org/10.1007/978-3-319-78613-1_1

1

The electron and ion Hamiltonians H_e and H_i include also the interaction with external (given) electromagnetic fields, while the interaction parts (electron-electron H_{ee}, ion-ion H_{ii} and electron-ion H_{ei}) are understood as pure Coulomb ones.

Of course, one cannot ignore the spin as a supplementary degree of motion. Sometimes, it is nevertheless compelling to include other effects of higher order in the inverse light velocity $\frac{1}{c}$. Even more, in the treatment of magnetic properties one has to redefine the model by including also the localized electrons of atomic cores. Furthermore, it is plausible, that for the understanding of some important phenomena one reaches the limitations of today's solid state theory.

An essential role in the mathematical treatment of this system plays the thermodynamic limit i.e. letting the number of particles and the volume of the system tend to infinity, while keeping the density of particles constant. The very existence of this limit for interacting quantum mechanical particles is not at all obvious, but necessary for the stability of matter.

To treat such a still extremely complicated system it is necessary to make further simplifications. Since most solids are crystals, one admits that only the valence electrons are allowed to move over the whole crystal, while the ions at most oscillate around their equilibrium positions in the given lattice. In a first step, one starts from the model of rigid ions (their mass is thousands of times heavier than the electron mass!) in a given periodical lattice and considers the motion of the electrons in a periodic field. Actually the characteristics of the lattice should be also determined by the above Hamiltonian, but one springs over this step. The task is still too complicated, and one treats not the many electron system, but just one electron in the field of ions and the self-consistent field of the other electrons. In a first step, one considers this potential as a given one. This oversimplified picture is already able to describe qualitatively the fundamental properties of solids. This is the *one-electron theory* of solid state, which will be described in Chap. 2 of this compendium. We make here also a first step toward the many-body treatment by considering many, but non-interacting electrons within the second quantization scheme in metals and semiconductors. In Chap. 3 we consider electron-electron interactions and many-body approximation schemes. Lattice oscillations and their interaction with the electron are discussed in Chap. 4. An important part of traditional solid-state theory concerns transport and optical properties. These will be presented respectively in Chaps. 5 and 6. We describe there also some new aspects related to the interaction with strong ultra-short laser pulses. Phase transformation, however important they are, we shall touch only in a few simple examples in Chap. 7. As a single modern subject we discuss in Chap. 8 some exotic properties of two-dimensional semiconductor structures. The Compendium ends with two Appendices. In Chap. 9 we give an overview of the concepts of theoretical physics we use. It is only a reminder and it is supposed that the reader is familiar with all of them. Further, in Chap. 10 some home-works are proposed for the interested reader.

Chapter 2
Non-interacting Electrons

Most properties of a crystal may be interpreted as the quantum mechanical motion of a single electron in a given potential. After the simplest cases of motion in homogeneous electric and magnetic fields with emphasis on the thermodynamic limit, an extended treatment of the motion in a periodical potential is given. The Bloch oscillations observed in periodical semiconductor layers are also included. The ground state occupation of the one-electron states leads to the understanding of the different classes of materials, as metals, insulators and semiconductors. The properties of the latter are strongly influenced by the presence of impurities. The controlled presence of donors and acceptors determine the properties of semiconductor contacts discussed on the example of a p-n contact.

2.1 Free Electrons

In the frame of quantum mechanics, the stationary state (wave function) of a free electron in the whole space is described by a plane wave

$$\psi(\mathbf{x}) = e^{i\mathbf{k}\mathbf{x}}$$

having a continuous energy spectrum (kinetic energy) depending only on $k = |\mathbf{k}|$

$$\varepsilon(\mathbf{k}) = \frac{\hbar^2}{2m}k^2 \qquad (0 < k < \infty) \, .$$

© Springer International Publishing AG, part of Springer Nature 2018
L. A. Bányai, *A Compendium of Solid State Theory*,
https://doi.org/10.1007/978-3-319-78613-1_2

This state is not normalized

$$\int d\mathbf{x} |\psi(\mathbf{x})|^2 = \infty$$

i.e. it is not a true eigenfunction.

Since we want to deal with true eigenfunctions and to consider systems with a definite volume (for simplicity a cube of size L^3) we may look for eigenfunctions in the product form $\psi(\mathbf{x}) = \phi_1(x_1)\phi_2(x_2)\phi_3(x_3)$ with two possible choices of boundary conditions in each dimension.

We may impose on the wave function $\phi(x)$ either to

(a) vanish at the boundary (Dirichlet): $\phi(\pm\frac{L}{2}) = 0$, then one gets symmetrical and anti-symmetrical eigenfuntions

$$\phi_k(x)^s = \sqrt{\frac{2}{L}} \cos(kx) \qquad \left(k = (2n+1)\frac{\pi}{L}; \quad n = 0, 1, 2, \cdots\right)$$

$$\phi_k(x)^a = \sqrt{\frac{2}{L}} \sin(kx) \qquad \left(k = n\frac{2\pi}{L}; \quad n = 1, 2, 3, \cdots\right)$$

or

(b) to be periodical: $\phi(x + L) = \phi(x)$, then the eigenfunctions are

$$\phi_k(x) = \frac{1}{\sqrt{L}} e^{\iota kx} \qquad \left(k = n\frac{2\pi}{L}; \quad n = \pm 1, \pm 2, \cdots\right).$$

The energy in both cases is given by $\varepsilon_k = \frac{\hbar^2}{2m}k^2$ however, with the above given discrete eigenvalues. The wave-functions $\phi_k(x)$ are normalized in the interval $[-\frac{L}{2}, \frac{L}{2}]$.

It is convenient to work with variant (b), since these eigenfunctions are also eigenfunctions of the momentum operator

$$-\iota\hbar\frac{\partial}{\partial x}\phi_k(x) = \hbar k\phi_k(x).$$

One has orthonormality (with the "bra" and "ket" symbols of Dirac)

$$< k|k' > \equiv \int_{-\frac{L}{2}}^{\frac{L}{2}} dx \phi_k(x)^* \phi_{k'}(x)$$

$$= \frac{1}{L}\int_{-\frac{L}{2}}^{\frac{L}{2}} dx e^{(k'-k)x} = \int_{-\frac{1}{2}}^{\frac{1}{2}} dy e^{2\pi(n'-n)y} = \delta_{k,k'}$$

as well as the completeness of Fourier series.

The electrons are fermions with spin $\frac{1}{2}$. The spin projection on the arbitrary chosen z-axis σ has two eigenstates with eigenvalues $\pm\frac{1}{2}$ and therefore the electron states ("kets") we have to consider are

$$|\mathbf{k}, \sigma >,$$

with the quantum numbers $\mathbf{k} = (k_1, k_2, k_3)$, $\sigma = \pm\frac{1}{2}$. However, the energy of the free electron is spin-independent.

If one wants to consider states with many electrons, then according to the Pauli principle for fermions one must build up anti-symmetrized products of one-electron wave functions. It is however advantageous to use the formalism of occupation numbers characterizing a many electron state by the occupation numbers of the one-electron states $\mathbf{k}, \sigma$. Due to the same Pauli principle these occupation numbers $n_{\mathbf{k},\sigma}$ may take only the values 0, and 1. This follows automatically from the anti-symmetry of the wave function in the configuration space. The total energy and total momentum of such a many electron state are

$$E = \sum_{\mathbf{k},\sigma} n_{\mathbf{k},\sigma} \varepsilon_{\mathbf{k},\sigma} \; ; \qquad \mathbf{P} = \sum_{\mathbf{k},\sigma} n_{\mathbf{k},\sigma} \hbar\mathbf{k}$$

with the occupation numbers and one-electron energies

$$n_{\mathbf{k},\sigma} = 0, 1 \; ; \qquad \varepsilon_{\mathbf{k},\sigma} = \frac{\hbar^2}{2m} k^2 ,$$

while the total number of electrons is

$$N = \sum_{\mathbf{k},\sigma} n_{\mathbf{k},\sigma} .$$

If one wants to study bulk properties of macroscopic matter (independent of the surface!), one has to perform the *thermodynamic limit*, that means in this case

$$L \to \infty , \qquad N \to \infty ,$$

while keeping constant the average density of electrons

$$\langle n \rangle = \frac{\langle N \rangle}{L^3} .$$

By this infinite volume limit procedure, the discrete sums go over into Riemann integrals. With $\Delta k_i = \frac{2\pi}{L}$ one has

$$\sum_{\mathbf{k}} \Delta\mathbf{k} \to \int d\mathbf{k} ,$$

or otherwise stated

$$\sum_{\mathbf{k}} \rightarrow \frac{L^3}{(2\pi)^3} \int d\mathbf{k} \ .$$

Since the one-electron energy is monotonously increasing with $k \equiv |\mathbf{k}|$, the ground state of a many-electron system is obviously the one with the one-electron states completely occupied up to a certain wave vector k_F

$$n_{\mathbf{k},\sigma} = \theta(k_F - |\mathbf{k}|) \ .$$

The energy of an electron with this wave-vector $\varepsilon_F \equiv \frac{\hbar^2}{2m}k_F^2$ is called the Fermi energy.

In thermodynamic equilibrium at a given temperature and chemical potential, described by the macro-canonical distribution the probability of a many electron state specified by the set of occupation numbers $\nu \equiv n_{\mathbf{k}_1,\sigma_1}, n_{\mathbf{k}_2,\sigma_2}, \ldots$ is given by

$$\mathscr{P}_\nu = \frac{e^{-\beta(E_\nu - \mu N_\nu)}}{\sum_{\nu'} e^{-\beta(E_{\nu'} - \mu N_{\nu'})}}$$

where $\beta = \frac{1}{k_B T}$, k_B is the Boltzmann constant, T is the absolute temperature and μ is the chemical potential. The average occupation number of an one-electron state $\mathbf{k}, \sigma$ is then

$$\langle n_{\mathbf{k},\sigma} \rangle = \sum_\nu \mathscr{P}_\nu n_{\mathbf{k},\sigma}$$

$$= \frac{\sum_{n=0,1} n e^{-\beta(\frac{\hbar^2}{2m}k^2 - \mu)n}}{\sum_{n=0,1} e^{-\beta(\frac{\hbar^2}{2m}k^2 - \mu)n}}$$

$$= \frac{1}{1 + e^{\beta(\frac{\hbar^2}{2m}k^2 - \mu)}} \equiv f(\frac{\hbar^2}{2m}k^2) \ ,$$

where

$$f(\varepsilon) = \frac{1}{1 + e^{\beta(\varepsilon - \mu)}}$$

is the Fermi function giving the average number of electrons in a state with energy ε. The total average number of particles is

$$\sum_{\sigma = \pm\frac{1}{2}} \sum_{\mathbf{k}} f(\frac{\hbar^2}{2m}k^2) = \langle N \rangle$$

and in the thermodynamic limit (after dividing with the volume L^3 of the system)

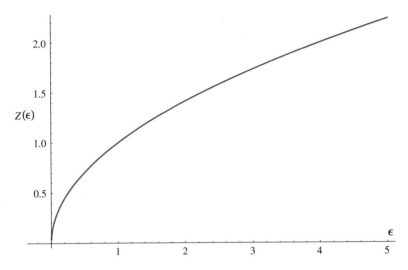

Fig. 2.1 Free one-electron state density $z(\varepsilon)$

$$\frac{2}{(2\pi)^3} \int d\mathbf{k}\, f(\frac{\hbar^2}{2m}k^2) = \langle n \rangle .$$

This equation determines the chemical potential μ at a given temperature T and average density $\langle n \rangle$. By introducing the one-electron state density

$$z(\varepsilon) \equiv \lim_{L\to\infty} \frac{1}{L^3} \sum_{\mathbf{k},\sigma} \delta(\varepsilon - \frac{\hbar^2}{2m}k^2) = \frac{2}{(2\pi)^3} \int d\mathbf{k}\,\delta(\varepsilon - \frac{\hbar^2}{2m}k^2) = \frac{1}{2\pi^2\hbar^3}(2m)^{\frac{3}{2}}\varepsilon^{\frac{1}{2}}$$

(shown in Fig. 2.1) it may be rewritten as

$$\int d\varepsilon\, z(\varepsilon) f(\varepsilon) = \langle n \rangle .$$

The average electron energy is

$$\langle \varepsilon \rangle = \frac{1}{\langle n \rangle} \int d\varepsilon\, z(\varepsilon) f(\varepsilon) \varepsilon .$$

One distinguishes two regimes:

(i) *degenerate*, with a big positive chemical potential $\beta\mu \gg 1$, which in the extreme case at $T = 0$ gives rise to a step-like Fermi function

$$f(\varepsilon)|_{T=0} = \theta(\varepsilon_F - \varepsilon) ,$$

the Fermi energy ε_F being defined by $\mu|_{T=0K} = \varepsilon_F > 0$ and the average one-electron energy is

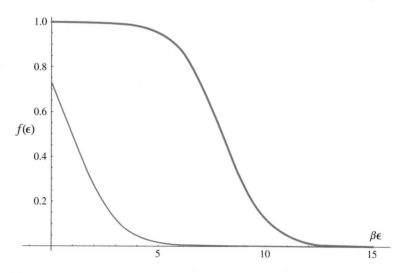

Fig. 2.2 Degenerate with $\beta\mu = 8$ (magenta) and non-degenerate with $\beta\mu = 1$ (blue) Fermi functions

$$\langle\varepsilon\rangle|_{T=0} = \frac{3}{5}\varepsilon_F \ .$$

(ii) *non-degenerate*, with $\beta\mu \leqslant 1$, which in the extreme case of a negative chemical potential with $\beta\mu \ll -1$ leads to the Boltzmann distribution

$$f(\varepsilon) \approx e^{\frac{\mu-\varepsilon}{k_B T}}$$

and the average one-electron energy

$$\langle\varepsilon\rangle \approx \frac{3k_B T}{2} \ .$$

An illustration of the shape of degenerate and non-degenerate Fermi functions is given in Fig. 2.2.

2.2 Electron in Electric and Magnetic Fields

The non-relativistic Hamiltonian of an electron in the presence of given (external) electromagnetic fields is

$$H = \frac{1}{2m}\left(\mathbf{p} + \frac{e}{c}\mathbf{A}\right)^2 - eV \ .$$

The scalar and vector-potentials V, $\mathbf{A}$ are related to the electric and magnetic fields by

$$\mathbf{E} = -\nabla V - \frac{1}{c}\dot{\mathbf{A}} \quad ; \qquad \mathbf{B} = \nabla \times \mathbf{A} .$$

It is important to remark, that the first term in the Hamiltonian is actually the usual kinetic energy since $\dot{\mathbf{r}} = \frac{1}{m}\left(\mathbf{p} + \frac{e}{c}\mathbf{A}\right)$.

2.2.1 Homogeneous, Constant Electric Field

A homogeneous, constant electric field may be described by different choices of the potentials, for example,

$$V(\mathbf{r}) = -\mathbf{r}\mathbf{E} \quad and \quad \mathbf{A} = 0$$

or

$$V(\mathbf{r}) = 0 \quad and \quad \mathbf{A} = -c\mathbf{E}t .$$

In the second choice one gets a time-dependent Hamiltonian. Due to gauge invariance, the physical results are independent of the choice, but it is rather convenient to work in the first gauge with the time-independent Hamiltonian

$$H = -\frac{\hbar^2}{2m}\nabla^2 + e\mathbf{r}\mathbf{E} .$$

Then, with $\mathbf{E} = (E, 0, 0)$ only the motion along the x-axis is affected by the electric field while the transverse motion is still described by plane waves. The solutions of the stationary Schroedinger equation with the energy $-\infty < \varepsilon < \infty$

$$\left(-\frac{\hbar^2}{2m}\frac{d^2}{dx^2} + eEx - \varepsilon\right)\phi_\varepsilon(x) = 0$$

are the Airy functions (seen in Fig. 2.3)

$$\phi_\varepsilon(x) = \frac{1}{\sqrt{\pi}}\int_0^\infty dq \cos\left(\frac{1}{3}q^3 + q\xi\right); \quad \xi \equiv \left(x - \frac{\varepsilon}{eE}\right)\left(\frac{2mE}{\hbar^2}\right)^{\frac{1}{3}} .$$

Although in this gauge we found stationary states it makes no sense to look for equilibrium since the energy spectrum is unbounded from below and these states are not eigenfunctions in the usual sense since they are not normalized

$$\int_{-\infty}^\infty dx |\phi_\varepsilon(x)|^2 = \infty$$

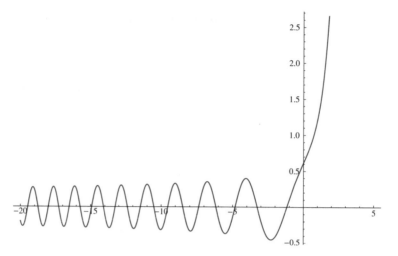

Fig. 2.3 Airy function

and the energy spectrum is continuous. This is related to the fact, that already a classical electron is accelerated in a homogeneous, constant electric field. This aspect we may recover also in the quantum mechanical theory if we write the above Hamiltonian in the momentum representation:

$$H(\mathbf{p}) = \frac{\mathbf{p}^2}{2m} - \iota e\hbar\mathbf{E}\frac{\partial}{\partial\mathbf{p}} \quad .$$

Then the Heisenberg equation of motion for the momentum

$$\dot{\mathbf{p}} = e\mathbf{E}$$

follows.

2.2.2 Homogeneous, Constant Magnetic Field

Let us choose the homogeneous, constant magnetic field along the z -axis

$$\mathbf{B} = (0, 0, B)$$

and the vector potential in the Landau gauge $\mathbf{A} = (0, Bx, 0)$. Then the Hamiltonian of the electron looks as

$$H = -\frac{\hbar^2}{2m}\frac{\partial^2}{\partial x^2} + \frac{1}{2m}\left(-\imath\hbar\frac{\partial}{\partial y} - \frac{e}{c}Bx\right)^2 - \frac{\hbar^2}{2m}\frac{\partial^2}{\partial z^2} \ .$$

One looks for the stationary solution of the Schroedinger equation in the form

$$\psi_{k_y,k_z}(\mathbf{r}) = \Phi(x)\frac{e^{\imath(k_z z + k_y y)}}{\sqrt{L_z L_y}} \ .$$

Then $\Phi(x)$ has to satisfy the equation

$$\left[-\frac{\hbar^2}{2m}\frac{\partial^2}{\partial x^2} + \frac{1}{2}m\omega_c^2(x - X)^2 - \varepsilon + \frac{\hbar^2 k_z^2}{2m}\right]\Phi(x) = 0 \ .$$

As usual, one chooses periodical boundary conditions along the y and z -axes. There-fore, $k_z = \frac{2\pi n_z}{L_z}$, $k_y = \frac{2\pi n_y}{L_y}$ with integer n_x, n_y. Here

$$\omega_c = \frac{|e|B}{mc}$$

is the frequency of the cyclotron oscillation,

$$X = sign(e)\ell_B^2 k_y$$

is the x coordinate of the center of the cyclotron motion, while

$$\ell_B = \sqrt{\frac{\hbar c}{|e|B}}$$

is the "magnetic length". As it may be seen, the problem is not completely sepa-rable, in the sense, that the motion along the x-axis depends also on the quantum numbers of the motion along the other axes. The normalized (on the whole x-axis!) eigenfunctions of this Schroedinger equation are given by

$$\Phi_{n,X}(x) = \frac{1}{\sqrt{\ell_B}}e^{-x^2/2\ell_B^2} H_n((x - X)/\ell_B)$$

where $H_n(x)$ are the Hermite polynomials and the corresponding eigenenergies are quantized (oscillator values), but degenerate with respect to X

$$\varepsilon_{n,X,k_z,} = \frac{\hbar^2 k_z^2}{2m} + \hbar\omega_c(n + \frac{1}{2}) \qquad (n = 0, 1, 2, ...) \ .$$

The oscillator ground-state eigenfunction $\Phi_{0,0}(x)$ with ℓ_B taken as unit length is shown in Fig. 2.4. In the x-direction the boundary is not fixed. One might however

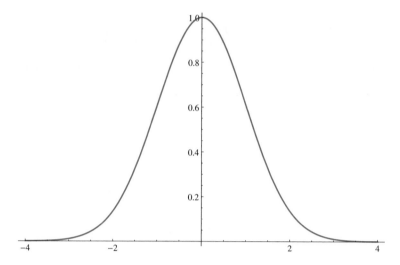

Fig. 2.4 The ground-state Landau function $\Phi_{0,0}(x)$

restrict the coordinate X to a certain domain $-L_x/2 \leq X \leq L_x/2$ in order to get approximate boundaries in the sense, that at distances much bigger than the magnetic length far away the wave function is very small. Typical for these Landau-functions is, that the energy does not depend either on the quantum number X nor on k_x. However, if one imposes Dirichlet (vanishing) boundary conditions at $\pm L_x/2$, the first degeneracy is lifted.

Another convenient way to impose a "confinement" along the x axis is to introduce an oscillator potential barrier

$$V(x) = \frac{m}{2}\omega_0^2 x^2 \ .$$

The solution is again the usual Landau function, however the cyclotron frequency gets replaced by

$$\tilde{\omega} = \sqrt{\omega_c^2 + \omega_0^2}$$

and the cyclotron center coordinate X gets replaced by

$$\tilde{X} = \frac{\omega_c^2}{\tilde{\omega}^2}X \ .$$

In this way, the degeneracy with respect to X is lifted and the eigenenergies are given by

$$\varepsilon_{n,\tilde{X},k_z} = \frac{\hbar^2 k_z^2}{2m} + \hbar\tilde{\omega}(n + \frac{1}{2}) + \frac{m\omega_0^2}{2}\tilde{X}^2 \qquad (n = 0, 1, 2, ...) \ .$$

Nevertheless, whenever effects at the boundaries do not play an essential role, it is mostly convenient to work with the Landau functions.

In the presence of the magnetic field it is also very important to take into account that one has an additional spin-dependent energy $-2\sigma\mu_B B$ where $\mu_B = \frac{e\hbar}{2mc}$ is the Bohr magneton and the projection of the spin on the z axis may take two values $\sigma = \pm\frac{1}{2}$.

2.2.2.1 Magnetization

Since in the case of the magnetic field, by restricting the cyclotron center as we mentioned before, we may construct normalized eigenfunctions with discrete spectrum bounded from below and we may look for equilibrium properties. For this sake let us consider the one-electron state density, where we took into account that the spin of the electron along the magnetic field may take two possible values $\sigma = \pm\frac{1}{2}$:

$$z(E) = \frac{1}{L_x L_y L_z} \sum_{k_z,k_y,n,\sigma} \delta(E - \frac{\hbar^2 k_z^2}{2m} - \hbar\omega_c(n + \frac{1}{2}) + s_z 2\mu_B B) \quad .$$

By performing the thermodynamic limit we have to take into account that by limiting the range of the center of the cyclotronic motion X implicitly $|k_y| \leq \frac{L_x}{2\ell_B^2}$. Then we get in the thermodynamic limit

$$z(E) = \frac{1}{(2\pi)^2 \ell_B^2} \sqrt{\frac{2m}{\hbar^2}} \sum_{n,\sigma}^{\infty} \frac{\theta(E - \hbar\omega_c(n + \frac{1}{2}) + \sigma 2\mu_B B)}{\sqrt{E - \hbar\omega_c(n + \frac{1}{2}) + \sigma 2\mu_B B}} \quad .$$

Obviously, one has a singular behavior at certain discrete periodical values of the energy E (see Fig. 2.5, where the unit of energy was taken to be $\hbar\omega_c$ and a red line shows the state density in the absence of the magnetic field). For sake of simplicity, the spin is ignored in the figure. Anyway, at very low temperatures all the spins are oriented along the magnetic field. This gives rise to typical oscillations of the equilibrium magnetization as function of $\frac{1}{B}$ (de Haas van Alfven effect).

Let us consider now a many-electron system in equilibrium. The equilibrium magnetization is defined as the derivative of the free energy F with respect to the magnetic field

$$M = -\frac{1}{L_x L_y L_z} \frac{\partial F}{\partial B} \quad .$$

The free energy is related on its turn to the thermodynamic potential $\mathscr{F}$ and the average number of particles $\langle N \rangle$ by

$$F = \mathscr{F} + \mu\langle N \rangle .$$

Fig. 2.5 State density $z(E)$
in magnetic field

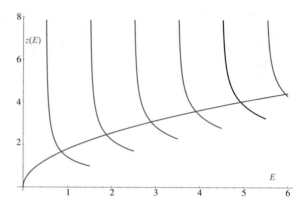

In the macro-canonical equilibrium, we have

$$\mathscr{F} = -k_B T \ln \left\{ Sp \left(e^{-\beta(H-\mu N)} \right) \right\} .$$

With the short-hand notation i for the one-electron Landau state quantum numbers
and $n_i = 0, 1$ for this occupation number we get

$$
\begin{aligned}
\mathscr{F} &= -k_B T \ln \left\{ \sum_{n_1, n_2, \dots} e^{-\beta \sum_i (\varepsilon_i - \mu) n_i} \right\} \\
&= -k_B T \ln \left\{ \Pi_i \left[1 + e^{-\beta(\varepsilon_i - \mu)} \right] \right\} \\
&= -k_B T \sum_i \ln \left[1 + e^{-\beta(\varepsilon_i - \mu)} \right] \\
&= -k_B T L_x L_y L_z \int d\varepsilon z(\varepsilon) \ln \left[1 + e^{-\beta(\varepsilon - \mu)} \right] .
\end{aligned}
$$

The average electron density $\langle n \rangle$ on its turn is defined by the one-electron state density
and the average occupation number of the states given by the Fermi function

$$f(\varepsilon) \equiv \frac{1}{1 + e^{\beta(\varepsilon - \mu)}} \quad ,$$

$$\langle n \rangle = \int d\varepsilon z(\varepsilon) f(\varepsilon)$$

and

$$F = \int d\varepsilon z(\varepsilon) \left(-k_B T \ln \left[1 + e^{-\beta(\varepsilon - \mu)} \right] + \mu \frac{1}{1 + e^{\beta(\varepsilon - \mu)}} \right) .$$

Thus, the magnetic field dependence of the one-particle state density (at a fixed
chemical potential μ) determines the magnetization.

2.2.3 Motion in An One-Dimensional Potential Well

Let us consider here a simple one-dimensional problem of quantum-mechanical motion in a static potential well. We have discussed already, how the energy spectrum of the free propagating particle can be conveniently discretized by Dirichlet boundary conditions at $x = \pm\frac{L}{2}$. Actually, this corresponds to the motion in an infinitely high potential well, that does not allow the penetration inside the potential barrier. The purpose was, to recuperate later the thermodynamic limit at $L \to \infty$ while keeping the average particle density constant. Now we shall consider here a potential well of finite height U_0 and in order to avoid confusions with the previous problem we denote here the finite width of the well with a.

The stationary Schroedinger equation we consider is

$$\left\{-\frac{\hbar^2}{2m}\frac{d^2}{dx^2} - eV(x)\right\}\psi(x) = E\psi(x) ,$$

with the potential

$$-eV(x) = \begin{cases} 0 & for \quad |x| \le \frac{a}{2} \\ U_0 > 0 & for \quad |x| > \frac{a}{2} \end{cases}$$

shown in Fig. 2.6.

The solutions inside the well are trigonometric functions $\sin(kx)$ and $\cos(kx)$ with $k^2 = \frac{2m}{\hbar^2}E$, while outside the well, choosing vanishing conditions at $x \to \pm\infty$ these are respectively e^{kx} for $x < 0$ and e^{-kx} for $x > 0$, with $k^2 = \frac{2m}{\hbar^2}|U_0 - E|$. Imposing the continuity of the wave function $\psi(x)$ and its derivative $\psi'(x)$ at $x = \pm\frac{a}{2}$, as well as the normalization condition $\int_{-\infty}^{\infty}|\psi(x)|^2 dx = 1$ one gets the transcendent equations for the eigenvalues of the even and odd solutions

$$\tan(ka) = \frac{\sqrt{C^2 - (ka)^2}}{ka} ,$$

Fig. 2.6 Potential well

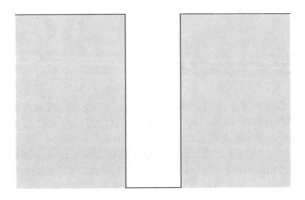

Fig. 2.7 Energy levels in the well for $C^2 = 38$

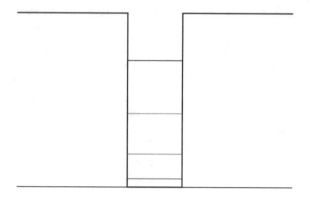

respectively

$$\tan(ka) = -\frac{ka}{\sqrt{C^2 - (ka)^2}} \ ,$$

where

$$C^2 = \frac{a^2 2m}{\hbar^2} U_0 \ .$$

The resulting energy spectrum consists of a finite number of discrete energies below U_0. For $C^2 = 38$ they are illustrated in Fig. 2.7. The shown levels correspond alternately to symmetric and anti-symmetric states. Of course, one has always a continuum above the barrier.

2.3 Electrons in a Periodical Potential

Periodical potentials in a crystal play a central role. The main properties of solids may be understood starting from the peculiarities of the quantum mechanical electron motion in a periodical potential.

2.3.1 Crystal Lattice

Most of the solid-state devices are made out of crystals and a theoretical treatment of periodical structures is much easier, than that of disordered structures. In what follows we shall discuss only such solids with a periodical structure.

A crystal is characterized first of all by a translation symmetry of the (average) positions $\mathbf{R}$ of the atoms. Such a symmetry is defined by a 3D Bravais lattice

$$\mathbf{R} = n_1\mathbf{a}_1 + n_2\mathbf{a}_2 + n_3\mathbf{a}_3; \qquad (n_1, n_2, n_3 = 0, \pm1, \pm2, \ldots) \ .$$

The three vectors $\mathbf{a}_1, \mathbf{a}_2, \mathbf{a}_3$ define the elementary cell, whose repetition covers all the atoms. Its choice is not unique, but should contain the minimal possible number of atoms. An illustration of a Bravais lattice with a single atom in the elementary cell is shown in Fig. 2.8, for sake of simplicity in 2D. An example of a more complicated elementary cell with two atoms is given on Fig. 2.9. One illustrates other possible choices of the elementary cell (see the arrows!). The volume of the elementary cell is given by

$$v = \mathbf{a}_1 \cdot (\mathbf{a}_2 \times \mathbf{a}_3)$$

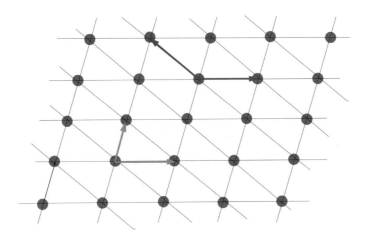

Fig. 2.8 Illustration of a crystal lattice in 2D, with two different choices of the elementary cell (indicated by the green, respectively blue arrows)

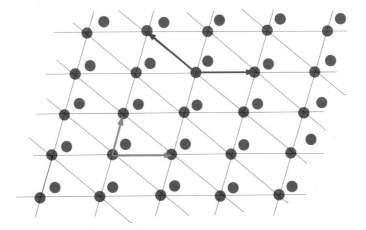

Fig. 2.9 Illustration of a Bravais lattice with an elementary cell of two atoms in 2D

and does not depend on the choice of the elementary cell. The elementary cell may contain different atoms and there are also crystals with elementary cells containing hundreds of atoms.

The discrete translational symmetry is not the only possible symmetry of a crystal. It may be symmetric also with respect to certain rotations or mirrorings. In what follows we shall consider only the periodicity, which determines the basic properties of any crystal. The unit vectors $\mathbf{a}_1, \mathbf{a}_2, \mathbf{a}_3$ are not orthogonal to each other (except in a cubic crystal), therefore it is useful to introduce also a dual basis (called "reciprocal") by

$$\mathbf{b}_1 = \frac{2\pi}{v} \mathbf{a}_2 \times \mathbf{a}_3$$

$$\mathbf{b}_2 = \frac{2\pi}{v} \mathbf{a}_3 \times \mathbf{a}_1$$

$$\mathbf{b}_3 = \frac{2\pi}{v} \mathbf{a}_1 \times \mathbf{a}_2 .$$

Then we have the orthogonality of the two dual unit vector sets:

$$\mathbf{b}_i \cdot \mathbf{a}_j = 2\pi \delta_{i,j} .$$

One may define also a reciprocal lattice created by the new basis vectors. A vector $\mathbf{K}$ of this reciprocal lattice is defined by

$$\mathbf{K} = m_1\mathbf{b}_1 + m_2\mathbf{b}_2 + m_3\mathbf{b}_3; \qquad (m_1, m_2, m_3 = 0, \pm1, \pm2, \dots) .$$

The volume of the elementary cell of this lattice v_{BZ} called Brillouin Zone (BZ) is related to that of the elementary cell of the original lattice v by

$$v_{BZ} = \mathbf{b}_1 \cdot (\mathbf{b}_2 \times \mathbf{b}_3) = \frac{(2\pi)^3}{\mathbf{a}_1 \cdot (\mathbf{a}_2 \times \mathbf{a}_3)} = \frac{(2\pi)^3}{v} .$$

The following relations also hold

$$\mathbf{R} \cdot \mathbf{K} = 2\pi (n_1 m_1 + n_2 m_2 + n_3 m_3)$$

$$e^{i\mathbf{R}\cdot\mathbf{K}} = 1$$

$$\frac{1}{v} \int_v d\mathbf{r} e^{i\mathbf{r}\cdot\mathbf{K}} = \delta_{\mathbf{K},0} .$$

Any periodical function f on the lattice ($f(\mathbf{r}) = f(\mathbf{r} + \mathbf{R})$) may be expanded in Fourier series

$$f(\mathbf{r}) = \sum_{\mathbf{K}} \tilde{f}_{\mathbf{K}} e^{i\mathbf{r}\cdot\mathbf{K}}$$

with the Fourier coefficients

$$\tilde{f}_{\mathbf{K}} = \frac{1}{v} \int_v d\mathbf{r} f(\mathbf{r}) e^{-i\mathbf{r}\cdot\mathbf{K}} .$$

2.3.2 Bloch Functions

Let us now consider the Hamiltonian of an electron

$$H = -\frac{\hbar^2}{2m}\nabla^2 + U(\mathbf{r})$$

in a periodical potential

$$U(\mathbf{r}) = U(\mathbf{r} + \mathbf{R}) .$$

Here we want to exploit just those general features that emerge from the periodicity. The stationary states are solutions of the equation

$$H\psi = E\psi .$$

Since a perfectly periodical crystal has an infinite extension, the stationary solutions (like the plane waves in free space) cannot be normalized, and in this sense, they are not true eigenstates. From the periodicity it follows, that the shifted wave function $\psi(\mathbf{r} + \mathbf{R})$ must be a stationary state with the same energy E and (in the absence of degeneracy) it may differ from the unshifted one only by a phase factor

$$\psi_E(\mathbf{r} + \mathbf{R}) = e^{i\phi_{\mathbf{R}}}\psi_E(\mathbf{r}) .$$

It follows immediately the additivity of the phase

$$\phi_{\mathbf{R}_1+\mathbf{R}_2} = \phi_{\mathbf{R}_1} + \phi_{\mathbf{R}_2}$$

and therefore one may write

$$\psi_E(\mathbf{r} + \mathbf{R}) = e^{i\mathbf{k}\mathbf{R}}\psi_E(\mathbf{r})$$

with a real vector $\mathbf{k}$. According to the previous definition of the reciprocal lattice, $\mathbf{k}$ and $\mathbf{k} + \mathbf{K}$ are equivalent, since the scalar product $\mathbf{R}\mathbf{K}$ is a multiple of 2π. One may thus write the wave function in the form

$$\psi_{\mathbf{k}}(\mathbf{r}) = e^{i\mathbf{k}\mathbf{r}}u_{\mathbf{k}}(\mathbf{r})$$

where $u_{\mathbf{k}}(\mathbf{r})$ is periodical

$$u_{\mathbf{k}}(\mathbf{r} + \mathbf{R}) = u_{\mathbf{k}}(\mathbf{r})$$

and the wave vector $\mathbf{k}$ belongs to the elementary cell of the reciprocal lattice. We choose it as the cell at the origin (BZ) defined as the domain

$$\mathbf{k}^2 \leq (\mathbf{k} - \mathbf{K})^2$$

of $\mathbf{k}$-vectors closer to the origin as any non-vanishing vector $\mathbf{K}$ of the reciprocal lattice. Even in the case of degeneracy one may choose the wave functions in this Bloch form. (This stems from the fact that the translations are an abelian group with one-dimensional irreducible representations.)

Thus, beside the energy we have the wave vector $\mathbf{k}$ for the characterization of the Bloch states. Of course, other quantum numbers must be considered, which we shall denote by n. These quantum numbers are discrete and are called–band indices.

2.3.3 Periodical Boundary Conditions

Just like in the case of the free electrons it is comfortable to impose some boundary conditions to get a discrete spectrum. In the case of a crystal lattice one chooses a periodical boundary condition compatible with the lattice periodicity

$$\psi(\mathbf{r} + \mathcal{N}_i \mathbf{a}_i) = \psi(\mathbf{r}).$$

with some integer $\mathcal{N}_i$ ($i = 1, 2, 3$). The parallelepiped with the edges $\mathcal{N}_i \mathbf{a}_i$ is called basic volume Ω. From the definition, it follows that the condition

$$\mathcal{N}_i \mathbf{a}_i \cdot \mathbf{k} = 2\pi m ; \qquad (m = 0, \pm 1, \pm 2, \ldots)$$

has to be satisfied in order to recover the same phase factor of the Bloch function and therefore the wave vector $\mathbf{k}$ may take only discrete values

$$\mathbf{k} = \frac{n_1}{\mathcal{N}_1} \mathbf{b}_1 + \frac{n_2}{\mathcal{N}_2} \mathbf{b}_2 + \frac{n_3}{\mathcal{N}_3} \mathbf{b}_3 ; \qquad (n_1, n_2, n_3 = 0, \pm 1, \pm 2, \ldots) .$$

Of course, to obtain physically relevant results for a bulk crystal one has to remove the dependence on these artificial boundary conditions by taking the infinite volume limit. One sees that as $\mathcal{N}_i$ ($i = 1, 2, 3$) increases, the elementary step of the wave vectors $\mathbf{k}$ goes to zero, therefore in the limit of infinite basic volume in all directions one has

$$\frac{1}{\mathcal{N}_1 \mathcal{N}_2 \mathcal{N}_3 v} \sum_{\mathbf{k} \in BZ} \longrightarrow \frac{1}{(2\pi)^3} \int_{BZ} d^3 k .$$

The introduction of the periodical boundary conditions allows the orthonormalization of the Bloch functions

$$\int_{\Omega} d\mathbf{r} \psi_{\mathbf{k},n}(\mathbf{r})^* \psi_{\mathbf{k}',n'}(\mathbf{r}) = \delta_{\mathbf{k},\mathbf{k}'} \delta_{n,n'} \ .$$

Now, the integration over the basic volume Ω is equivalent to a summation over all the elementary cells and the integration over each cell. With

$$\psi_{\mathbf{k},n}(\mathbf{r}) = \frac{1}{\sqrt{\mathcal{N}}} e^{i\mathbf{k}\mathbf{r}} u_{\mathbf{k},n} \ ; \qquad \mathcal{N} \equiv \mathcal{N}_1 \mathcal{N}_2 \mathcal{N}_3$$

the orthonormality may be written also as

$$\frac{1}{\mathcal{N}} \sum_{\mathbf{R} \in \Omega} e^{i(\mathbf{k}-\mathbf{k}')\mathbf{R}} \int_{v} d\mathbf{r} e^{i(\mathbf{k}-\mathbf{k}')\mathbf{r}} u_{\mathbf{k},n}(\mathbf{r})^* u_{\mathbf{k}',n'}(\mathbf{r}) = \delta_{\mathbf{k},\mathbf{k}'} \int_{v} d\mathbf{r} u_{\mathbf{k},n}(\mathbf{r})^* u_{\mathbf{k}',n'}(\mathbf{r}) \ ,$$

where we used the periodicity of the Bloch part and the relation

$$\frac{1}{\mathcal{N}} \sum_{\mathbf{R} \in \Omega} e^{i\mathbf{k}\mathbf{R}} = \delta_{\mathbf{k},0}$$

stemming from the identity

$$\frac{1}{\mathcal{M}} \sum_{n=0}^{\mathcal{M}-1} x^n = \frac{1}{\mathcal{M}} \frac{1-x^{\mathcal{M}}}{1-x} = \begin{cases} 0 \ if \ m \neq 0 \\ 1 \ if \ m = 0 \end{cases} ,$$

with $x \equiv e^{i\frac{2\pi m}{\mathcal{M}}}$; $(m = 0, \ldots, \mathcal{M}-1)$. Thus,

$$\int_{v} d\mathbf{r} u_{\mathbf{k},n}(\mathbf{r})^* u_{\mathbf{k},n'}(\mathbf{r}) = \delta_{n,n'} \ .$$

and therefore the Bloch parts $u_{\mathbf{k},n}(\mathbf{r})$ form for each $\mathbf{k}$ an orthonormalized set and at any fixed $\mathbf{k}$ one has a discrete spectrum.

2.3.4 The Approximation of Quasi-Free Electrons

Let us write the Schroedinger eigenvalue equation in the periodical potential $U(\mathbf{r})$ in Fourier components

$$u_{\mathbf{k}}(\mathbf{r}) = \sum_{\mathbf{K}} \tilde{u}_{\mathbf{k},\mathbf{K}} e^{i\mathbf{K}\cdot\mathbf{r}}$$

$$U(\mathbf{r}) = \sum_{\mathbf{K}} \tilde{U}_{\mathbf{K}} e^{i\mathbf{K}\cdot\mathbf{r}} \quad,$$

where $\tilde{U}_{\mathbf{K}} = \tilde{U}^*_{-\mathbf{K}}$ follows from reality of the potential and the wave vector $\mathbf{k}$ here is restricted to the Brillouin zone. After multiplication with $e^{-i\mathbf{K}\cdot\mathbf{r}}$ and integration of $\mathbf{r}$ over the elementary cell the equation looks as

$$\left[\frac{\hbar^2}{2m}(\mathbf{k}+\mathbf{K})^2 - E_{\mathbf{K}}(\mathbf{k}) \right] \tilde{u}_{\mathbf{k},\mathbf{K}} + \sum_{\mathbf{K}'} \tilde{U}_{\mathbf{K}-\mathbf{K}'} \tilde{u}_{\mathbf{k},\mathbf{K}'} = 0 \,.$$

Let us consider first an empty lattice $U(\mathbf{r}) = 0$. The corresponding states and eigenenergies are

$$\psi^{(0)}_{\mathbf{k},\mathbf{K}}(\mathbf{r}) = \frac{1}{\sqrt{\mathcal{N}}} e^{i\mathbf{k}\mathbf{r}} u^{(0)}_{\mathbf{k},\mathbf{K}}(\mathbf{r}), \qquad u^{(0)}_{\mathbf{k},\mathbf{K}}(\mathbf{r}) = \frac{1}{\sqrt{v}} e^{i\mathbf{K}\mathbf{r}} \quad,$$

respectively

$$E^{(0)}_{\mathbf{K}}(\mathbf{k}) = \frac{\hbar^2}{2m}(\mathbf{k}+\mathbf{K})^2; \qquad \mathbf{k} \in BZ \,.$$

The "bands" indicated by the vectors $\mathbf{K}$ of the reciprocal lattice are shown in Fig. 2.10 for $K = -2\pi, 0, 2\pi$, (with lattice constant $a = 1$) in the case of an one-dimensional lattice.

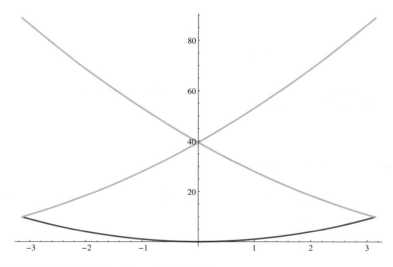

Fig. 2.10 The first folded free electron bands in 1D

It is important to remark again, that at fixed k the spectrum is discrete. One sees the degeneracy of the "bands" at $k = \pm\pi$ and at $k = 0$. The main role of the perturbation by a small periodical potential $U(\mathbf{r})$ will be in lifting these degeneracies. As we shall see on an exactly soluble one-dimensional example, after the lifting of the degeneracy the top of the lowest band and the bottom of the upper band get extrema with vanishing derivative.

2.3.5 The Kronig-Penney Model

We consider here explicitly the solvable one-dimensional periodical potential model (see Fig. 2.11)

$$V(x) = \begin{cases} 0 & n(a+b) < x < n(a+b)+a; \quad n = 0, \pm 1, \pm 2, \ldots \\ U_0 & n(a+b)-b \leq x \leq n(a+b) . \end{cases}$$

Actually one has to solve the problem in the elementary cell $(0 < x < l, \equiv a+b)$ and thereafter extend it overall by the Bloch condition. For $E < U_0$ the solution is

$$\Psi(x) = c_1^{II} e^{i\kappa_2 x} + c_2^{II} e^{-i\kappa_2 x}; \qquad 0 < x < a$$
$$\Psi(x) = c_1^{I} e^{\kappa_1 x} + c_2^{I} e^{-\kappa_1 x}; \qquad a < x < l$$

$$\kappa_1^2 = \frac{2m}{\hbar^2} E, \qquad \kappa_2^2 = \frac{2m}{\hbar^2} |U_0 - E| .$$

At first one must ensure the continuity of the wave function and its derivative at $x = a$

$$\Psi(a-0) = \Psi(a+0)$$
$$\Psi'(a-0) = \Psi'(a+0) .$$

Fig. 2.11 The Kronig-Penney periodic potential

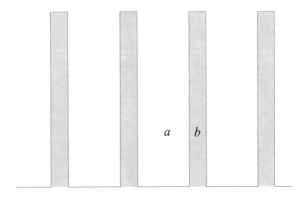

Further one has still to fulfill the Bloch condition

$$\Psi(0)e^{\imath kl} = \Psi(l)$$
$$\Psi'(0)e^{\imath kl} = \Psi'(l).$$

These conditions lead to a system of four linear homogeneous equations for the four coefficients c_1^I, c_1^{II}, c_2^I, c_2^{II}. The existence of a solution implies a transcendent equation for the energy as function of the wave vector k (energy bands). For $E < U_0$ it looks as

$$\cosh \kappa_2 b \cos \kappa_1 a + \frac{\kappa_2^2 - \kappa_1^2}{2\kappa_1\kappa_2} \sinh \kappa_2 b \sin \kappa_1 a = \cos k(a+b).$$

Fig. 2.12 The lowest four energy bands (here $\frac{2ma^2}{\hbar^2} E$) of the Kronig-Penney model with
$b = 0.25a$, $\frac{2ma^2}{\hbar^2} U_0 = 38$
for $0 < k(a+b) < \pi$

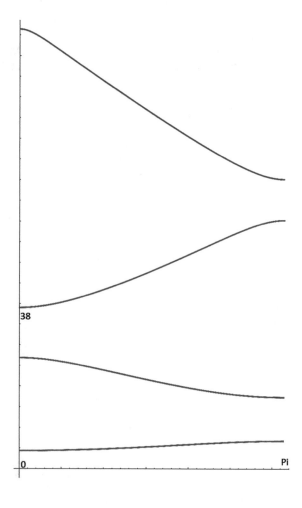

Fig. 2.13 The lowest four energy bands (here $\frac{2ma^2}{\hbar^2}E$) of the Kronig-Penney model with $b = a$, $\frac{2ma^2}{\hbar^2}U_0 = 38$ for $0 < k(a + b) < \pi$. (Bands below U_0 here in red)

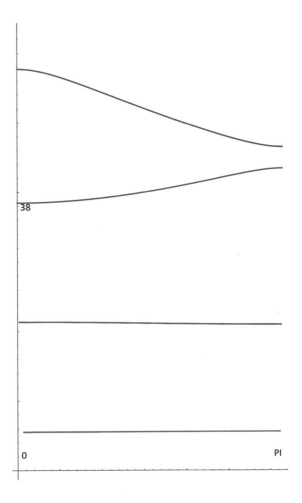

For $E > U_0$ trigonometric functions are the solution and the above discussed boundary conditions lead then to the transcendent equation

$$\cos \kappa_1 b \cos \kappa_2 a - \frac{\kappa_1^2 + \kappa_2^2}{2\kappa_1 \kappa_2} \sin \kappa_1 b \sin \kappa_2 a = \cos k(a + b)$$

where however

$$\kappa_1^2 = \frac{2m}{\hbar^2} E, \qquad \kappa_2^2 = \frac{2m}{\hbar^2}|E - U_0|.$$

These equations may be easily solved numerically. The lowest four bands are shown in Fig. 2.12 by the choice of the parameters $U_0 = 38$, $b = 0.25a$.

One sees here the scenario described in the previous section. The bands of the empty lattice are split, and one gets extrema with vanishing derivatives.

It is instructive to remark here, that as the distance between the wells b increases the bands under U_0 flatten (in our case the lowest two) and get close to the energy levels of the isolated well, while the upper bands get closer to those of the empty lattice. This is illustrated in Fig. 2.13 with the increased distance $b = a$ between the wells. (The range of the wave vector is limited here for sake of convenience to $0 < k(a + b) < \pi$.) This result corresponds to the known fact that far away atoms have the spectrum of the isolated atoms, however, with a degeneracy equaling the total number of atoms.

2.3.6 Band Extrema, kp–Perturbation Theory and Effective Mass

The stationary Schroedinger equation for the Bloch functions describing the energy bands (indexed by some quantum numbers n)

$$\left[-\frac{\hbar^2}{2m}\nabla^2 + U(\mathbf{r}) - E_n(\mathbf{k}) \right] \psi_{n,\mathbf{k}}(\mathbf{r}) = 0$$

after inserting

$$\psi_{n,\mathbf{k}}(\mathbf{r}) = e^{\iota \mathbf{k}\mathbf{r}} u_{n,\mathbf{k}}(\mathbf{r});$$

gives rise to the Bloch equation

$$\left[-\frac{\hbar^2}{2m}\nabla^2 + U(\mathbf{r}) - \frac{\iota\hbar^2}{m}\mathbf{k}\nabla + \frac{\hbar^2}{2m}k^2 - E_n(\mathbf{k}) \right] u_{\mathbf{k},n}(\mathbf{r}) = 0.$$

For any fixed $\mathbf{k}$ this is a true eigenvalue equation with discrete eigenvalues. The eigenfunctions $u_{\mathbf{k},n}(\mathbf{r})$ due to the Bloch property

$$u_{n,\mathbf{k}}(\mathbf{r} + \mathbf{R}) = u_{n,\mathbf{k}}(\mathbf{r})$$

may be considered just in the elementary cell with periodical boundary conditions. Therefore, they may form a complete orthonormalized set in the cell

$$\int_v d\mathbf{r} u_{n,\mathbf{k}}(\mathbf{r})^* u_{n',\mathbf{k}}(\mathbf{r}) = \delta_{n,n'}$$

$$\sum_n u_{n,\mathbf{k}}(\mathbf{r}) u_{n,\mathbf{k}}(\mathbf{r}')^* = \delta(\mathbf{r}, \mathbf{r}').$$

Let us assume, that for a given $\mathbf{k}_0$ the eigenfunctions $u_{n,\mathbf{k}_0}(\mathbf{r})$ and eigenenergies $E_n(\mathbf{k}_0)$ are known.

For a small deviation $|\mathbf{k} - \mathbf{k}_0| \to 0$, one may consider $-\frac{\imath\hbar^2}{m}(\mathbf{k} - \mathbf{k}_0)\nabla$ as a small perturbation. In the absence of degeneration, standard perturbation theory gives

$$u_{n,\mathbf{k}}(\mathbf{r}) = u_{n,\mathbf{k}_0}(\mathbf{r}) + \sum_{n' \neq n} \frac{<n', \mathbf{k}_0| \frac{\hbar^2(\mathbf{k}-\mathbf{k}_0)\nabla}{\imath m}|n, \mathbf{k}_0>}{E_n(\mathbf{k}_0) - E_{n'}(\mathbf{k}_0)} u_{n',\mathbf{k}_0}(\mathbf{r}) + \cdots$$

and

$$E_n(\mathbf{k}) - \frac{\hbar^2}{2m}k^2 = E_n(\mathbf{k}_0) - \frac{\hbar^2}{2m}k_0^2 + <n, \mathbf{k}_0|\frac{\hbar(\mathbf{k}-\mathbf{k}_0)}{\imath m}\nabla|n, \mathbf{k}_0>$$

$$+ \sum_{n' \neq n} \frac{\left| <n, \mathbf{k}_0|\frac{\hbar^2(\mathbf{k}-\mathbf{k}_0)\nabla}{\imath m}|n', \mathbf{k}_0> \right|^2}{E_n(\mathbf{k}_0) - E_{n'}(\mathbf{k}_0)} + \cdots$$

with the "bra-ket" notation

$$<n, \mathbf{k}_0|\nabla|n', \mathbf{k}_0> \equiv \int_v d\mathbf{r} u_{n,\mathbf{k}_0}(\mathbf{r})^* \nabla u_{n',\mathbf{k}_0}(\mathbf{r}).$$

It follows then for the derivative of the energy at $\mathbf{k} = \mathbf{k}_0$

$$\frac{\partial E_n(\mathbf{k})}{\partial \mathbf{k}}|_{\mathbf{k}=\mathbf{k}_0} = <n, \mathbf{k}_0|\frac{\hbar^2}{\imath m}\nabla|n, \mathbf{k}_0>.$$

Since $\mathbf{k}_0$ for the time being is arbitrary, the relation holds for any $\mathbf{k}$ and we get for the average velocity

$$\langle \mathbf{v} \rangle \equiv \int_v d\mathbf{r} \psi_{n,\mathbf{k}}(\mathbf{r})^* \frac{\hbar}{\imath m}\nabla \psi_{n,\mathbf{k}}(\mathbf{r}) = \frac{1}{\hbar}\frac{\partial E_n(\mathbf{k})}{\partial \mathbf{k}}.$$

Now let us assume, that the point $\mathbf{k}_0$ is a band extremum. Then the linear terms in the energy vanish and

$$E_n(\mathbf{k}) = E_n(\mathbf{k}_0) + \frac{\hbar^2}{2}\left(\frac{1}{\mathscr{M}}\right)_{\mu v}(k - k_0)_\mu(k - k_0)_v + \cdots$$

where

$$\left(\frac{1}{\mathscr{M}}\right)_{\mu v} = \frac{1}{m}\delta_{\mu v} + \frac{2}{m^2}\sum_{n' \neq n} \frac{<n, 0|\hbar\nabla_\mu|n', 0><n', 0|\hbar\nabla_v|n, 0>}{E_n(0) - E_{n'}(0)} + \cdots$$

is the *inverse effective mass tensor.* In the case of the cubic symmetry with the extremum at $\mathbf{k}_0 = 0$ this looks as

$$E_n(\mathbf{k}) = E_n(0) + \frac{\hbar^2}{2m^*}k^2 + \cdots$$

and one has an isotropic *effective mass* m^*. This effective mass may be positive or negative, depending on the nature of the extremum (maximum or minimum). Therefore, depending on the sign of the effective mass, the average velocity

$$\langle \mathbf{v} \rangle = \frac{\hbar}{m^*}\mathbf{k}$$

may be either in the direction of the wave vector $\mathbf{k}$ or opposite to it.

We have seen on the example of the one-dimensional Kronig-Penney model, that band extrema occur at $k = 0, \pm\frac{\pi}{a}$. Most of the important properties of crystals are determined by the extrema of some of the energy bands we shall discuss later.

2.3.7 Wannier Functions and Tight-Binding Approximation

A useful concept in the treatment of motion in a periodical potential are the Wannier functions defined through the Bloch functions of a given band n by

$$w_{n\mathbf{R}}(\mathbf{r}) \equiv \frac{1}{\sqrt{\mathcal{N}}} \sum_{\mathbf{k} \in BZ} \psi_{n,\mathbf{k}}(\mathbf{r})e^{-i\mathbf{k}\mathbf{R}} = \frac{1}{\sqrt{\mathcal{N}}} \sum_{\mathbf{k} \in BZ} u_{n,\mathbf{k}}(\mathbf{r})e^{i\mathbf{k}(\mathbf{r}-\mathbf{R})} .$$

They constitute a a complete orthonormalized system of functions

$$\int_\Omega d\mathbf{r} w_{n\mathbf{R}}(\mathbf{r})^* w_{n'\mathbf{R}'}(\mathbf{r}) = \frac{1}{\mathcal{N}} \sum_{\mathbf{k},\mathbf{k}' \in BZ} e^{i(\mathbf{k}\mathbf{R}-\mathbf{k}'\mathbf{R}')} \int_v d\mathbf{r}\psi_{n,\mathbf{k}}(\mathbf{r})^* \psi_{n',\mathbf{k}'}(\mathbf{r})$$

$$= \frac{1}{\mathcal{N}} \sum_{\mathbf{k} \in BZ} e^{i\mathbf{k}(\mathbf{R}-\mathbf{R}')}\delta_{n,n'} = \delta_{n,n'}\delta_{\mathbf{R},\mathbf{R}'}$$

and

$$\sum_{n,\mathbf{R}} w_{n\mathbf{R}}(\mathbf{r})^* w_{n\mathbf{R}}(\mathbf{r}') = \delta(\mathbf{r},\mathbf{r}') .$$

The inverse transformation is

$$u_{n,\mathbf{k}}(\mathbf{r}) = \frac{1}{\sqrt{\mathcal{N}}} \sum_{\mathbf{R}} e^{i\mathbf{k}(\mathbf{r}-\mathbf{R})} w_{n,\mathbf{R}}(\mathbf{r}) .$$

The Wannier functions however are not eigenfunctions of the Hamiltonian. If $u_{n,\mathbf{k}}(\mathbf{r})$ is a smooth function of $\mathbf{k}$, then $w_{n\mathbf{R}}(\mathbf{r})$ vanishes rapidly for $|\mathbf{r} - \mathbf{R}| \rightarrow \infty$. Therefore, while the Bloch eigenfunctions are delocalized, the Wannier functions are localized on the lattice nodes. Since the Wannier functions are constructed only from the Bloch functions of a single band and considering also the discrete translational invariance on the lattice, the matrix elements of the Hamiltonian between the Wannier functions depend only on the vector $\mathbf{R} - \mathbf{R}'$

$$< n, \mathbf{R}|H|n', \mathbf{R}' >= \delta_{n,n'} t_n (\mathbf{R} - \mathbf{R}').$$

Therefore,

$$
\begin{aligned}
< n, \mathbf{R}|H|n, \mathbf{k} > &= \frac{1}{\sqrt{\mathcal{N}}} \sum_{\mathbf{R}'} < n, \mathbf{R}|H|n, \mathbf{R}' > e^{-i\mathbf{k}\mathbf{R}'} \\
&= \frac{1}{\sqrt{\mathcal{N}}} \sum_{\mathbf{R}'} t_n (\mathbf{R} - \mathbf{R}') e^{i\mathbf{k}\mathbf{R}'} \\
&= e^{i\mathbf{k}\mathbf{R}} \frac{1}{\sqrt{\mathcal{N}}} \sum_{\mathbf{R}'} t_n (\mathbf{R}') e^{i\mathbf{k}\mathbf{R}'}.
\end{aligned}
$$

On the other hand

$$
\begin{aligned}
< n, \mathbf{R}|H|n, \mathbf{k} > &= E_n(\mathbf{k}) < n, \mathbf{R}|n, \mathbf{k} > \\
&= E_n(\mathbf{k}) \frac{1}{\sqrt{\mathcal{N}}} \sum_{\mathbf{R}'} < n, \mathbf{R}|n, \mathbf{R}' > e^{i\mathbf{k}\mathbf{R}'}. \\
&= e^{i\mathbf{k}\mathbf{R}} E_n(\mathbf{k}) \frac{1}{\sqrt{\mathcal{N}}}
\end{aligned}
$$

and therefore, inverting the relation we get

$$E_n(\mathbf{k}) = \sum_{\mathbf{R}} e^{i\mathbf{k}\mathbf{R}} t_n (\mathbf{R}).$$

As an illustration, let us assume that in a cubic lattice of lattice constant a the only non-vanishing $t_n(\mathbf{R})$ are those to the nearest neighbors

$$t_n(0) = c_n, \quad t_n(\pm \mathbf{a}_x) = t_n(\pm \mathbf{a}_y) = t_n(\pm \mathbf{a}_z) = -d_n.$$

Then it follows,

$$E_n(\mathbf{k}) = c_n - 2d_n(\cos(k_x a) + \cos(k_y a) + \cos(k_z a))$$

and in the vicinity of the origin

$$E_n(\mathbf{k}) = c_n - 6d_n + d_n a^2 k^2 + \dots$$

This defines the effective mass by $\frac{\hbar^2}{2m^*} = d_n a^2$.

One approximates often the Wannier functions by the eigenfunctions of the isolated atom $\phi_n(\mathbf{r})$ ("tight-binding approximation")

$$w_{n\mathbf{R}}(\mathbf{r}) \approx \phi_n(\mathbf{r} - \mathbf{R}) .$$

Within this approximation

$$\psi_{n,\mathbf{k}}(\mathbf{r}) = \frac{1}{\sqrt{\mathcal{N}}} \sum_{\mathbf{R}} e^{-i\mathbf{k}\mathbf{R}} \phi_n(\mathbf{r} - \mathbf{R})$$

or

$$u_{n,\mathbf{k}}(\mathbf{r}) = \frac{1}{\sqrt{\mathcal{N}}} \sum_{\mathbf{R}} e^{i\mathbf{k}(\mathbf{r} - \mathbf{R})} \phi_n(\mathbf{r} - \mathbf{R}) .$$

It is however important to remark that the orthonormalization of these functions is not fulfilled since $\int_{\Omega} d\mathbf{r} \phi_n(\mathbf{r})^* \phi_{n'}(\mathbf{r} - \mathbf{R})$, however small, it does not vanish for $\mathbf{R} \neq 0$. By this connection to the atomic wave functions one may speak about a given band as characterized by a given symmetry of the underlying atomic states. For example, s or p-like states, with their respective degeneracies.

2.3.8 Bloch Electron in a Homogeneous Electric Field

Let us consider a homogeneous electric field superimposed on the periodical potential. Thus, we have a supplementary term $H' = -e\mathbf{E}\mathbf{r}$ in the Hamiltonian. The resulting potential looks like in Fig. 2.14. Let us choose the normalization of the Bloch functions as

$$< \mathbf{k}', n' | \mathbf{k}, n > = v_{BZ}^{-1} \delta(\mathbf{k} - \mathbf{k}') \delta_{n,n'} .$$

Since in the infinite volume limit $\mathcal{N} \delta_{\mathbf{k},\mathbf{k}'} \to v_{BZ} \delta(\mathbf{k} - \mathbf{k}')$, this normalization corresponds in the discretized version to $< \mathbf{k}', n' | \mathbf{k}, n > = N \delta_{\mathbf{k},\mathbf{k}'} \delta_{n,n'}$. Then

$$\int d\mathbf{r} \psi_{\mathbf{k}',n'}(\mathbf{r})^* \mathbf{r} \psi_{\mathbf{k},n}(\mathbf{r}) = \int d\mathbf{r} e^{i(\mathbf{k}-\mathbf{k}')\mathbf{r}} \mathbf{r} u_{\mathbf{k}',n'}(\mathbf{r})^* u_{\mathbf{k}\,n}(\mathbf{r})$$

$$= -i \frac{\partial}{\partial \mathbf{k}} \int d\mathbf{r} \psi_{\mathbf{k}',n'}(\mathbf{r})^* \psi_{\mathbf{k},n}(\mathbf{r}) + i \int d\mathbf{r} e^{i(\mathbf{k}-\mathbf{k}')\mathbf{r}} u_{\mathbf{k}',n'}(\mathbf{r})^* \frac{\partial}{\partial \mathbf{k}} u_{\mathbf{k},n}(\mathbf{r})$$

$$= -i v_{BZ}^{-1} \frac{\partial}{\partial \mathbf{k}} \delta(\mathbf{k} - \mathbf{k}') \delta_{n,n'} + i \sum_{\mathbf{R}} e^{i(\mathbf{k}-\mathbf{k}')\mathbf{R}} \int_v d\mathbf{r} e^{i(\mathbf{k}-\mathbf{k}')\mathbf{r}} u_{\mathbf{k}',n'}(\mathbf{r})^* \frac{\partial}{\partial \mathbf{k}} u_{\mathbf{k},n}(\mathbf{r})$$

$$= -i v_{BZ}^{-1} \frac{\partial}{\partial \mathbf{k}} \delta(\mathbf{k} - \mathbf{k}') \delta_{n,n'} + i v_{BZ} \delta(\mathbf{k} - \mathbf{k}') \int_v d\mathbf{r} u_{\mathbf{k},n'}(\mathbf{r})^* \frac{\partial}{\partial \mathbf{k}} u_{\mathbf{k},n}(\mathbf{r}) .$$

If one ignores inter-band matrix elements as well as the intra-band energy correction we get the approximate expression for the matrix elements of H':

$$< \mathbf{k}', n' | H' | \mathbf{k}\, n > \approx -i v_{BZ}^{-1} e\mathbf{E} \frac{\partial}{\partial \mathbf{k}} \delta(\mathbf{k} - \mathbf{k}') \delta_{n,n'} .$$

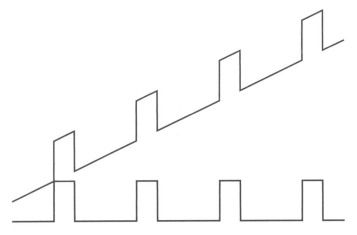

Fig. 2.14 A periodic potential and its modification due to a homogeneous electric field

Consequently, we obtain in the **k** space an effective one-band Hamiltonian (in each band)

$$H'_{eff} \equiv E(\mathbf{k}) - \iota e\mathbf{E}\frac{\partial}{\partial \mathbf{k}}$$

and in the same representation the following Heisenberg equation of motion for the **k**-vector

$$\hbar\dot{\mathbf{k}} = e\mathbf{E}$$

results.

Due to the periodicity of the band energy $E(\mathbf{k})$, after a time $\mathscr{T} = \frac{2\pi\hbar}{ea\mathscr{E}}$ the electron crosses the whole Brillouin zone and will be reflected at its boundary. Therefore, unlike the accelerated motion of a free electron, the electron in a crystal performs Bloch oscillations with the frequency $\omega = \frac{2\pi}{\mathscr{T}}$. However, these oscillations in a real crystal are not observable since due to the smallness of the lattice constant the Bloch period $\mathscr{T}$ is much bigger than the relaxation time due to the interaction with other perturbations of the lattice (phonons, impurities). The Bloch period itself cannot be shortened by the increase of the field strength due to the here neglected inter-band transitions, which may become important in strong electric fields. Nevertheless, Bloch oscillation may be observed in artificial periodical semiconductor structures with large lattice constants.

As in the case of the free electrons in a homogeneous field we may reinterpret these results in terms of stationary states (Wannier-Ladders). The stationary Schroedinger equation within our approximations looks as

$$\left[E(\mathbf{k}) - \iota e\mathbf{E}\frac{\partial}{\partial \mathbf{k}} - \varepsilon\right]\tilde{\psi}(\mathbf{k}) = 0$$

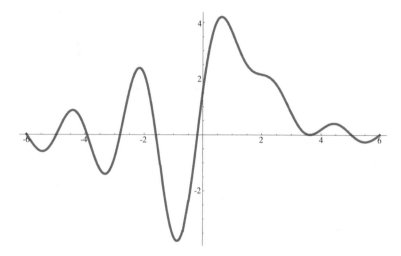

Fig. 2.15 Stationary state of the Wannier ladder

or

$$\frac{\mathbf{E}}{\mathcal{E}}\frac{\partial \ln \tilde{\psi}(\mathbf{k})}{\partial \mathbf{k}} = \frac{\imath}{e\mathcal{E}}(\varepsilon - E(\mathbf{k})) \,.$$

Choosing the field along the x-axis we get

$$\tilde{\psi}(\mathbf{k}) = \tilde{\psi}(0, \mathbf{k}_{\perp}) \exp\left\{\frac{\imath}{e\mathcal{E}}\int_0^{k_x} dq_x\,(\varepsilon - E(q_x, \mathbf{k}_{\perp}))\right\}\,.$$

We must still implement the periodic boundary condition for the wave function in the $\mathbf{k}$ space. Taking into account the periodicity of band energy $E(\mathbf{k})$ (implying the vanishing of its integral over the Brillouin zone). This is achieved by fixing the eigenenergies

$$\varepsilon_m(\mathbf{k}_{\perp}) = ea\mathcal{E}m + const. \qquad (m = 0, \pm 1, \pm 2, \ldots)\,.$$

Then the difference between two successive energy levels is

$$\varepsilon_{m+1}(\mathbf{k}_{\perp}) - \varepsilon_m(\mathbf{k}_{\perp}) = ea\mathcal{E}\,.$$

It corresponds to the energy associated to the Bloch frequency $\omega = \frac{2\pi}{\mathscr{T}}$.

As an example, let us consider a simple one-dimensional Bloch spectrum $E(k) = \Delta(1 - \cos(ka))$. Then $\varepsilon_m = ea\mathcal{E}m + \Delta$ and the wave function in real space (illustrated in Fig. 2.15 for $m = 0$) is

$$\psi_m(x) = const. \int_{-\frac{\pi}{a}}^{\frac{\pi}{a}} dk \exp\left\{-\iota k(x - ma) - \frac{\iota \Delta}{e\mathscr{E}} \sin(ka)\right\}.$$

2.4   Electronic Occupation of States in a Crystal

The occupation of the one-electron states in the periodical potential by as many electrons as ions at low temperatures determines the nature of the solid.

2.4.1   Ground State Occupation of Bands. Conductors and Insulators

We have discussed until now the nature of states of an electron in a crystal. However, the properties of the material are essentially determined by the occupation of these states. According to their electric conductance one classifies the crystals as metals or insulators. On the basis of the discussed one-electron spectra of crystals one may already understand the basic difference. The ground state of the electrons ($T = 0$) in a crystal results from the successive occupation of the lowest one-electron states in the bands. In this ground state corresponding to zero temperature, the highest band may be either partially or completely filled. In the first case, a small external perturbation can excite the electrons, while in the latter case, due to the forbidden states above them (band gap), only a strong perturbation may excite them. The partially filled band is called conduction band, while the last completely filled band is called valence band. The first case defines a conductor (metal), while the second defines an insulator (dielectric). The two situations are illustrated in the simple case of a direct band gap, when both the maximum of the valence band and the minimum of the conduction band lie at $\mathbf{k} = 0$ on Fig. 2.16. The red color indicates the occupied states in the bands.

Fig. 2.16 Band filling in metals and insulators

However, there is a very important sort of dielectrics, the semiconductors. They behave as insulators at very low temperatures, while at higher temperatures they behave as conductors. In this case the band gap is so small, that at room temperatures already a lot of electrons are thermally excited into the conduction band. In the presence of impurity states in the band gap, we shall discuss later, the number of excited electrons in the conduction band is even more important, and the conductivity may be relevant even at low temperatures.

Beside the above discussed cases, by crystals with indirect gaps, where the maximum of the valence band and the minimum of the conduction band occur at different values of $\mathbf{k}$, it may happen, that the later lies higher than the former. This band occupation at $T = 0$ of a semi-metal is schematically shown in Fig. 2.17.

2.4.2 Spin–Orbit Coupling and Valence Band Splitting

The discussion of the band-structure of crystals is treated usually within the frame of the non-relativistic Schroedinger equation. However, there are often important corrections to be made due to relativistic effects. Most of the so called direct gap semiconductors, having the maximum of the valence band and the minimum of the conduction band at $\mathbf{k} = 0$, within this frame get a non-degenerate conduction band and a threefold degenerate valence band. In a tight-binding approximation these may be considered as made up of s, respectively p atomic states. Of course, one has to take into account also the spin degeneracy. However, actually the valence band states are split due to relativistic effect of the spin-orbit interaction. This interaction is invariant against simultaneous rotation of the orbital angular momentum $\mathbf{l}$ and of the spin $\boldsymbol{\sigma}$. Therefore, one has to classify the valence -band states according to the states of the total angular momentum $\mathbf{J} \equiv \mathbf{l} + \boldsymbol{\sigma}$. From the $m = 0, \pm 1$ orbital angular momentum states and the $\sigma_z = \pm\frac{1}{2}$ spin states one has to construct the eigenstates of the total angular momentum $J = \frac{3}{2}$, $J_z = \pm\frac{1}{2}, \pm\frac{3}{2}$ and $J = \frac{1}{2}$, $J_z = \pm\frac{1}{2}$

Fig. 2.17 Band filling in a semi-metal

Fig. 2.18 Split valence
bands

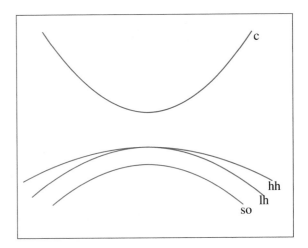

$$|\tfrac{3}{2}, \pm\tfrac{3}{2}> = \mp \qquad\qquad |1, \pm1 > |\tfrac{1}{2}, \pm\tfrac{1}{2} >$$

$$|\tfrac{3}{2}, \pm\tfrac{1}{2}> = \mp \tfrac{1}{\sqrt{3}} \left(|1, \pm1 > |\tfrac{1}{2}, \mp\tfrac{1}{2} > \mp \sqrt{2}|1, 0 > |\tfrac{1}{2}, \pm\tfrac{1}{2} > \right)$$

$$|\tfrac{1}{2}, \pm\tfrac{1}{2}> = \mp \tfrac{1}{\sqrt{3}} \left(\sqrt{2}|1, \pm1 > |\tfrac{1}{2}, \mp\tfrac{1}{2} > \pm |1, 0 > |\tfrac{1}{2}, \pm\tfrac{1}{2} > \right).$$

The phenomenological effective-mass Luttinger Hamiltonian for the valence
bands used in applications for crystals having the cubic symmetry

$$H = -\frac{\hbar^2}{2m_0} \left(\mathbf{p}^2 + m_1(\mathbf{pJ})^2 + \Delta J(J+1) - \frac{5}{4} \right)$$

describes three double degenerate valence bands around $\mathbf{k} = 0$. In Fig. 2.18 such a
split valence-band structure is shown schematically.

2.5   Electron States Due to Deviations from Periodicity

2.5.1   Effective Mass Approximation

Let us consider a deviation from the crystalline periodicity due to a non-periodical
potential $V(\mathbf{r})$ that varies slowly on the scale of the lattice constants. It may be due
to a defect in the lattice or a foreign atom. Then we may approximate

$$< n, \mathbf{R}|V|n', \mathbf{R}' > = \int d\mathbf{x} w_{n\mathbf{R}}(\mathbf{r})^* V(\mathbf{r}) w_{n'\mathbf{R}'}(\mathbf{r}) \approx V(\mathbf{R})\delta_{n,n'}\delta_{\mathbf{R},\mathbf{R}'} .$$

We may look for the eigenstates in the presence of both the periodical potential $U(\mathbf{r})$ and this potential $V(\mathbf{r})$ in terms of Wannier functions

$$\frac{1}{\sqrt{\mathscr{N}}} \sum_{\mathbf{R}} \chi(\mathbf{R}) w_{n\mathbf{R}}(\mathbf{r}) .$$

The eigenvalue equation then looks as

$$\sum_{\mathbf{R}'} \left(\tilde{E}(\mathbf{R} - \mathbf{R}') + (V(\mathbf{R}) - \varepsilon)\delta_{\mathbf{R},\mathbf{R}'} \right) \chi(\mathbf{R}') = 0$$

with

$$\tilde{E}_n(\mathbf{R}) \equiv t_n(\mathbf{R}) = \frac{1}{\mathscr{N}} \sum_{\mathbf{k}} E_n(\mathbf{k}) e^{\imath \mathbf{k}\mathbf{R}} .$$

On the other hand, if we introduce the interpolating function

$$\chi(\mathbf{r}) \equiv \sum_{\mathbf{k} \in BZ} e^{\imath \mathbf{k}\mathbf{r}} \tilde{\chi}(\mathbf{k})$$

for all $\mathbf{r}$, then

$$\sum_{\mathbf{R}'} \tilde{E}(\mathbf{R} - \mathbf{R}')\chi(\mathbf{R}') = \sum_{\mathbf{k}} E_n(\mathbf{k}) \tilde{\chi}(\mathbf{k}) e^{\imath \mathbf{k}\mathbf{R}}$$

$$= \sum_{\mathbf{k}} \tilde{\chi}(\mathbf{k}) E_n(-\imath \nabla) e^{\imath \mathbf{k}\mathbf{r}}|_{\mathbf{r}=\mathbf{R}} = E_n(-\imath \nabla)\chi(\mathbf{r})|_{\mathbf{r}=\mathbf{R}} .$$

Assuming that $\tilde{\chi}(\mathbf{k})$ vanishes rapidly far away from the band extremum at $\mathbf{k}_0$, where the expansion

$$E_n(\mathbf{k}) \approx E_n(\mathbf{k}_0) + \frac{\hbar^2}{2m^*}(\mathbf{k} - \mathbf{k}_0)^2 + \ldots$$

is valid and defining $\chi(\mathbf{r}) = e^{\imath \mathbf{k}_0 \mathbf{r}} \varphi(\mathbf{r})$ we get the Schroedinger equation with the effective mass m^*

$$\left\{ E_n(\mathbf{k}_0) - \frac{\hbar^2}{2m^*}\nabla^2 + V(\mathbf{r}) - \varepsilon \right\} \varphi(\mathbf{r}) = 0 .$$

Under the above assumptions we have also

$$\frac{1}{\sqrt{\mathscr{N}}} \sum_{\mathbf{R}} \chi(\mathbf{R}) w_{n\mathbf{R}}(\mathbf{r}) \approx \varphi(\mathbf{r}) \frac{1}{\sqrt{\mathscr{N}}} \sum_{\mathbf{R}} e^{\imath \mathbf{k}_0 \mathbf{R}} w_{n\mathbf{R}}(\mathbf{r}) = \varphi(\mathbf{r}) u_{n,\mathbf{k}_0}(\mathbf{r}) .$$

2.5.2 Intrinsic Semiconductors at Finite Temperatures

Since the importance of semiconductors in modern technological applications is overwhelming we devote more space to their discussion. Pure semiconductor materials, without impurities are called intrinsic. Here the last two relevant bands (valence and conduction bands) are neither completely filled or empty at ordinary temperatures. The electronic occupation of the Bloch-states with wave vector $\mathbf{k}$ and spin σ in equilibrium at inverse temperature β is described by the Fermi distribution

$$f_{n,\mathbf{k},\sigma} \equiv \langle n_{n,\mathbf{k},\sigma} \rangle = \frac{1}{e^{\beta(\varepsilon_{n,\mathbf{k},\sigma} - \mu)} + 1} .$$

It is meaningful to define the conduction band electrons as "true" electrons

$$f^e_{\mathbf{k},\sigma} \equiv f_{c,\mathbf{k},\sigma} = \frac{1}{e^{\beta(\varepsilon_{c,\mathbf{k},\sigma} - \mu)} + 1} ,$$

while in the valence band it is useful to define the population by the holes (absence of valence band electrons). With these definitions at $T = 0$ we have no holes and no electrons i.e. an electron-hole vacuum. As we shall see later, this terminology is extremely useful.

At the band edges of a direct gap semiconductor we have in the effective mass approximation the electron energy (in the conduction band) as

$$\varepsilon^e_{\mathbf{k}} = \varepsilon_{c,\mathbf{k},\sigma} = \frac{1}{2} E_g + \frac{\hbar^2}{2m^*_e} k^2 \quad (m^*_e > 0) ,$$

while we may define the energy of a hole (in the valence band) as

$$\varepsilon^h_{\mathbf{k}} = -\varepsilon_{v,-\mathbf{k},-\sigma} = \frac{1}{2} E_g + \frac{\hbar^2}{2m^*_h} k^2 \quad (m^*_h = -m^*_v > 0) .$$

This last definition corresponds to the fact that removing an electron of charge $e < 0$, spin σ, wave vector $\mathbf{k}$ and energy $\varepsilon_{v,\mathbf{k},\sigma}$ is equivalent to adding a particle with charge $-e > 0$, spin $-\sigma$, wave vector $-\mathbf{k}$ and energy $-\varepsilon_{v,-\mathbf{k},-\sigma}$. Thus, holes may be looked at as positively charged particles with positive effective mass. In a homogeneous electric field, they would be accelerated accordingly.

The average number of conduction electrons and holes is the same i.e. the system is electrically neutral

$$\sum_{\mathbf{k},\sigma} f^e_{\mathbf{k},\sigma} = \sum_{\mathbf{k},\sigma} f^h_{\mathbf{k},\sigma} .$$

If the electron-hole occupation of states far away of the band extrema is negligible one may extrapolate the effective mass approximation overall and we have in the thermodynamic limit

$$\frac{1}{(2\pi)^3} \int d\mathbf{k} \left[\frac{1}{e^{\beta(\frac{\hbar^2 k^2}{2m_e} + \frac{1}{2}E_g - \mu)} + 1} - \frac{1}{e^{\beta(\frac{\hbar^2 k^2}{2m_h} + \frac{1}{2}E_g + \mu)} + 1} \right] = 0 \ .$$

Since the band gap is still much bigger than the thermal energy ($\beta E_g \ll 1$) one may approximate the Fermi distribution through its non-degenerate limit and therefore

$$\frac{1}{(2\pi)^3} \int d\mathbf{k} \left[e^{-\beta(\frac{\hbar^2 k^2}{2m_e} + \frac{1}{2}E_g - \mu)} - e^{-\beta(\frac{\hbar^2 k^2}{2m_h} + \frac{1}{2}E_g + \mu)} \right] = 0 \ .$$

These integrals may be solved exactly and give rise to a chemical potential sitting in the gap at $\mu = \frac{3}{4}k_B T \ln\left(\frac{m_h}{m_e}\right)$ and the average number of each of the newly defined particles (carrier density) is

$$\langle n_0 \rangle = \sqrt{2\pi}(m_e m_h)^{\frac{3}{4}}(\beta\hbar^2)^{-\frac{3}{2}} e^{-\frac{1}{2}\beta E_g} \ .$$

2.5.3 Ionic Impurities

Let us see now what kind of modifications are introduced in the spectrum by putting an extra positive or negative ion in the crystal. Obviously, a minor modification of the continuous bands is not relevant. However, extra discrete states in the forbidden gap may be very important. Therefore, vicinities of band extrema must be looked up. Let us consider a positive ion in the origin. The stationary Schroedinger equation for an electron in the effective mass approximation in the neighborhood of a conduction band minimum is

$$\left\{ E_c - \frac{\hbar^2}{2m^*}\nabla^2 - \frac{e^2}{\varepsilon|\mathbf{r}|} - \varepsilon \right\} \varphi(\mathbf{r}) = 0 \qquad (m^* > 0) \ ,$$

where ε is the dielectric constant of the crystal and E_c denotes the energetic position of the band minimum. The ground state energy of this hydrogen-like Coulomb problem is well-known to be given by the Rydberg energy, here modified by the replacement of the electron mass by his effective mass m^* and the presence of the dielectric constant ε of the semiconductor. The binding energy here is the distance of the discrete level from the conduction band

$$\varepsilon - Ec = -E_R \qquad \left(E_R = \frac{\hbar^2}{2m^* a_B^2}; \quad a_B = \frac{\varepsilon\hbar^2}{m^* e^2} \right) \ .$$

The energy of this localized state therefore lies below the band minimum, also in the gap.

In the case of a negative ion placed at the origin we may expect analogously a new state in the band gap near a band maximum described by the equation

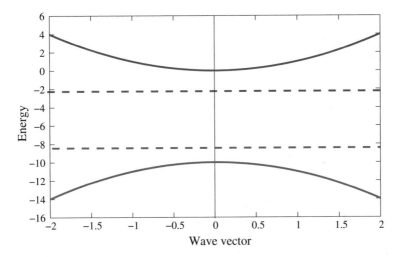

Fig. 2.19 Localized states in the energy gap of a semiconductor

$$\left\{ E_V - \frac{\hbar^2}{2m^*}\nabla^2 + \frac{e^2}{\varepsilon|\mathbf{r}|} - \varepsilon \right\} \varphi(\mathbf{r}) = 0 \qquad (m^* < 0)\,.$$

with a negative effective mass m^*. This may be rewritten as

$$\left\{ -E_V - \frac{\hbar^2}{2|m|^*}\nabla^2 - \frac{e^2}{\varepsilon|\mathbf{r}|} + \varepsilon \right\} \varphi(\mathbf{r}) = 0$$

giving rise to the ground state energy

$$\varepsilon - E_V = +E_R$$

lying in the gap above the band maximum. These states in the forbidden energy zone as they are illustrated in Fig. 2.19 are extremely important in the case of semiconductors.

In most of the semi-conducting crystals the dielectric constant is much greater than unity and the effective mass is much smaller than the true electron mass and, as a consequence, the Bohr Radius a_B is much bigger than the lattice constant. Thus, the assumption about the smoothness of the potential, as well as the use of the macroscopic dielectric constant ε of the material are justified. On the other hand, in a semiconductor the Rydberg energy is much smaller than the energy gap and of course even much smaller than the hydrogenic Rydberg energy.

2.5.4 Extrinsic Semiconductors at Finite Temperatures. Acceptors and Donors

As we already discussed, foreign ions in a semiconductor crystal lattice give rise to localized states in the band gap. On the other hand, one includes not ions, but neutral atoms. The question is, what happens if the atom gets ionized and losses or gains an electron in favor of the collectivized band electrons? A foreign atom may create a localized state just below the conduction band and remain neutral in the ground state. This would be interpreted as a positive ion state occupied by an electron. Another possibility is to create a localized state just above the valence band. Then in the ground state it will lose its electron in favor of the lower lying valence band continuum. This would be interpreted as a positive unoccupied ion. In the first case we speak about donors, while in the second about acceptors.

Bloch states with a given wave vector $\mathbf{k}$ may be occupied by two electrons of opposite spin. Double occupation of localized states however, may be forbidden. Indeed, the average Coulomb repulsion energy for two electrons on the same site is

$$U \equiv \frac{e^2}{\varepsilon} \int d\mathbf{r} \int d\mathbf{r}' \frac{|\psi_\uparrow(\mathbf{r})|^2 |\psi_\downarrow(\mathbf{r}')|^2}{|\mathbf{r} - \mathbf{r}'|}$$

In the case of the Coulomb ground state ($l = 0, \quad n = 1$)

$$\psi_\uparrow(\mathbf{r}) = \psi_\downarrow(\mathbf{r}) = \frac{1}{\sqrt{\pi a_B^3}} e^{-\frac{r}{a_B}}$$

the integral may be solved analytically giving rise to

$$U = \frac{5}{8} \frac{e^2}{\varepsilon a_B} = \frac{5}{4} E_R .$$

Therefore, the energy of the double occupied state may differ essentially from twice the energy of the single occupation. In this case one may write the energy of the donor state as in terms of its occupation by electrons as

$$E_D(n_\uparrow, n_\downarrow) = \varepsilon_D(n_\uparrow + n_\downarrow) + U n_\uparrow n_\downarrow ; \qquad (n_\uparrow, n_\downarrow = 0, 1) .$$

The average occupation at a finite temperature is given by

$$\langle n_\uparrow \rangle \equiv \langle n_\downarrow \rangle = \frac{\sum_{n_\uparrow=0,1} \sum_{n_\downarrow=0,1} n_\uparrow e^{-\beta[E_D(n_\uparrow,n_\downarrow)-\mu(n_\uparrow+n_\downarrow)]}}{\sum_{n_\uparrow=0,1} \sum_{n_\downarrow=0,1} e^{-\beta[E_D(n_\uparrow,n_\downarrow)-\mu(n_\uparrow+n_\downarrow)]}}$$

and

$$\langle n \rangle \equiv \langle n_\uparrow \rangle + \langle n_\downarrow \rangle = 2 \frac{e^{-\beta(\varepsilon_D-\mu)} + e^{-\beta[2(\varepsilon_D-\mu)+U]}}{1 + 2e^{-\beta(\varepsilon_D-\mu)} + e^{-\beta[2(\varepsilon_D-\mu)+U]}} \equiv \mathscr{F}_D$$

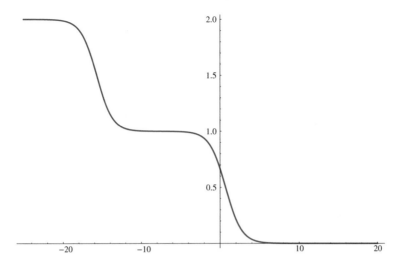

Fig. 2.20 The electron occupation $\mathscr{F}_D$ of a donor state

instead twice of the Fermi function one would have without repulsion. This new distribution is illustrated in Fig. 2.20

In the extreme case of very low temperatures $\frac{U}{k_B T} \to \infty$ it gives

$$\langle n \rangle = 2 \frac{e^{-\beta(\varepsilon_D - \mu)}}{1 + 2e^{-\beta(\varepsilon_D - \mu)}} = \frac{1}{e^{\beta(\varepsilon_D - \mu - k_B T \ln 2)} + 1} \; ,$$

which looks like a Fermi function with a shifted chemical potential

$$\mu \to \mu + k_B T \ln 2 \; .$$

Analogously we have for the positively charged holes on an acceptor level

$$E_A(m_\uparrow, m_\downarrow) = \varepsilon_A(m_\uparrow + m_\downarrow) + U m_\uparrow m_\downarrow \; ; \qquad (m_\uparrow, m_\downarrow = 0, 1)$$

with $m_\uparrow$, $m_\downarrow$ being the hole occupations. On has still to take into account that the chemical potential for the holes is $-\mu$ is and $\varepsilon_A = \frac{1}{2} E_g - E_R$.

The equation for the electro-neutrality in an extrinsic semiconductor looks therefore as

$$2 \sum_k f_k^e + \sum_D \mathscr{F}_D - N_D = 2 \sum_k f_k^h + \sum_A \mathscr{F}_A - N_A \; .$$

This equation may be solved for the chemical potential only numerically. As illustration, we show on Fig. 2.21 the one electron state density and its occupation in an n-type semiconductor with more donors (D) as acceptors (A) at $T = 0$. The green

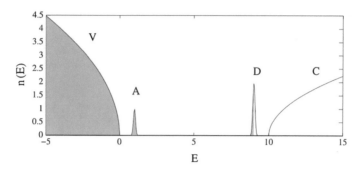

Fig. 2.21 State density in a compensated n-type semiconductor at $T = 0$

filling shows the (simple or double) occupation of the states. The donor states lying higher in energy lose some of their electrons in favor of the acceptor sates.

2.6 Semiconductor Contacts

Until now we discussed only bulk properties of solids, however phenomena at the interfaces play an overwhelming role in all solid-state devices. We describe in the following two such simple cases: the penetration of a static electric field in a semiconductor and the contact between an n-type extrinsic semiconductor and a p-type extrinsic semiconductor of the same crystal. In both cases thermal equilibrium is assumed.

A consequent quantum-mechanical discussion of these problems would be extremely difficult. A finite semiconductor is not a periodical crystal and the eigenstates and eigenenergies differ from that of the infinite one. A simplified quasi-classical approach however has proved to be very successful. We need to describe a situation in which the quantum-mechanical nature of the crystals, as well as macroscopic inhomogeneities must be reconciled. The basic object here to consider is the Wigner function depending on the momentum **p** and coordinate **x**. It is defined as

$$f(\mathbf{p}, \mathbf{x}, t) = \int d\mathbf{y}\, e^{\frac{i\mathbf{p}\mathbf{y}}{\hbar}} \langle \psi(\mathbf{x} + \frac{1}{2}\mathbf{y}, t)^+ \psi(\mathbf{x} - \frac{1}{2}\mathbf{y}, t) \rangle ,$$

where $\psi(\mathbf{x}, t)$ is the second quantized wave function and $\langle \ldots \rangle$ means averaging over a given ensemble. This implies the normalization

$$\int d\mathbf{x} \int \frac{d\mathbf{p}}{(2\pi\hbar)^3} f(\mathbf{p}, \mathbf{x}, t) = \langle N \rangle ,$$

where N is the operator of the total number of particles. It may be shown, that all the averages of operators which are a sum of two operators $O(\mathbf{x}, \frac{\hbar}{i}\nabla) \equiv A(\mathbf{x}) + B(\frac{\hbar}{i}\nabla)$

depending on the coordinate respectively on the momentum, may be expressed as the integrals

$$\langle O(\mathbf{x}, \frac{\hbar}{\iota}\nabla)\rangle = \int d\mathbf{x} \int \frac{d\mathbf{p}}{(2\pi\hbar)^3} f(\mathbf{p}, \mathbf{x}, t) O(\mathbf{x}, \mathbf{p}) .$$

However, generally speaking $f(\mathbf{p}, \mathbf{x})$ is neither real, nor positive. Nevertheless, in the quasi-classical limit ($\hbar \to 0$) it may be shown, that it will be real and positive. The Hamiltonian of a particle in the presence of a potential $U(\mathbf{x})$

$$H = -\frac{\hbar^2\nabla^2}{2m} + U(\mathbf{x})$$

has the above described structure.

One assumes tacitly, that the corresponding Wigner function in this quasi-classical limit, in thermal equilibrium is given again by the Fermi distribution

$$f(\mathbf{p}, \mathbf{x}) = \frac{1}{e^{\beta\left(\frac{p^2}{2m}+U(\mathbf{x})-\mu\right)} + 1} .$$

in the phase space $(\mathbf{x}, \mathbf{p})$.

2.6.1 Electric Field Penetration into a Semiconductor

In a previous discussion we have admitted implicitly, that a homogeneous constant electric field may be created in an infinite crystal. This is an extreme idealization. It is well-known, that metals screen out static electric fields, while in a semiconductor it may still penetrate at a finite macroscopic depth. We shall discuss here this last case within the quasi-classical approach. Let us consider the surface of an intrinsic semiconductor characterized by a dielectric constant ε and an equal number N of electrons and holes in the macroscopic volume Ω. Their density in equilibrium n_0 is determined by the band gap E_g and the temperature T and for $E_g \gg k_B T$ it is

$$n_0 \equiv N/\Omega \sim e^{-\frac{E_g}{2k_BT}} .$$

Let us consider, that the semiconductor occupies the half space $x > 0$. The static electric potential V(x) satisfies he Poisson equation outside

$$\nabla^2 V(\mathbf{x}) = 0 \quad (x < 0)$$

respectively inside the semiconductor

$$\varepsilon\nabla^2 V(\mathbf{x}) = -4\pi\rho(\mathbf{x}) \quad (x > 0) .$$

where $\rho(\mathbf{r})$ is the charge density of carriers (electrons and holes). In the presence of an external electric field $\mathscr{E}$ the potential outside the semiconductor is

$$V(\mathbf{x}) = -\mathscr{E}x + const. \qquad (x < 0) \ .$$

While in the absence of the field the charge density vanishes, in the presence of the field in equilibrium at a finite temperature it is determined self-consistently by the potential itself. In a quasi-classical approach (at high temperatures) the Maxwell-Boltzmann distribution gives the probability density of a state $\mathbf{r}, \mathbf{v}$ in the phase space

$$f(\mathbf{v}, \mathbf{x}) = \frac{e^{-\beta(\frac{mv^2}{2} + eV(\mathbf{x}))}}{\int d\mathbf{v}' \int d\mathbf{x}' e^{-\beta(\frac{mv^2}{2} + eV(\mathbf{x}'))}} \ .$$

This defines the equilibrium charge density of electrons and holes as

$$\rho_{e,h}(\mathbf{x}) = \pm eN \frac{e^{\mp \frac{eV(\mathbf{x})}{k_B T}}}{\int d\mathbf{r}' e^{\mp \frac{eV(\mathbf{x}')}{k_B T}}} \ .$$

Under the assumption, that the potential vanishes deep in the semiconductor, the integral in the denominator is proportional to the volume Ω and for a macroscopic volume we get approximately

$$\rho_{e,h}(\mathbf{x}) = \pm en_0 e^{\mp \frac{eV(\mathbf{x})}{k_B T}}$$

and for $eV \ll k_B T$ we have the charge density

$$\rho(\mathbf{x}) = \rho_e(\mathbf{x}) + \rho_h(\mathbf{x}) = -2en_0 \frac{eV(\mathbf{x})}{k_B T} \ .$$

Therefore, we get the self-consistent equation

$$\varepsilon \nabla^2 V(\mathbf{x}) = \frac{8\pi e^2 n_0}{k_B T} V(\mathbf{x})$$

for the determination of the potential inside the semiconductor. By introducing the Debye length

$$L_D \equiv \sqrt{\frac{k_B T \varepsilon}{8\pi e^2 n_0}}$$

the equation looks as

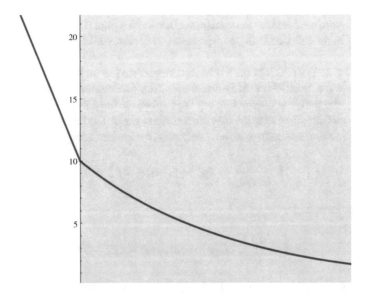

Fig. 2.22 Penetration of the potential into the semiconductor

$$\nabla^2 V(\mathbf{x}) - \frac{1}{L_D^2} V(\mathbf{x}) = 0$$

Now we must still consider the electromagnetic boundary condition

$$\frac{\partial V(x)}{\partial x}\Big|_{x=-0} = \varepsilon \frac{\partial V(x)}{\partial x}\Big|_{x=+0} \ .$$

to connect the solutions inside and outside. This gives rise to

$$V(x) = \mathscr{E}\frac{L_D}{\varepsilon} e^{-\frac{x}{L_D}}$$

for $x > 0$ and it is illustrated in Fig. 2.22.

   To conclude: Due to the screening by the carriers, the field penetrates at a depth L_D into the semiconductor. At room temperatures, this Debye length may be of order of centimeters and therefore our macroscopic treatment, as well as the existence of a more or less homogeneous electric field inside a semiconductor gets justified.

2.6.2   p-n Contact

Let us now consider the contact of two semiconductors of the same material, but one of them is p-doped (i.e. with acceptors), while the other one is n-doped (i.e. with donors). Again, we consider the contact surface to be a plane transverse to the x-axis at $x = 0$. On the left side is the p-type semiconductor, while on the right side the n-type.

Since we want to discuss here both non-degenerate as well as degenerate cases, we must use the quasi-classical Fermi distribution for fermions defined above.

So long the two semiconductors are kept apart, we have for the bulk materials a homogeneous situation with the electro-neutrality condition for the electrons, holes in the bands and the negative, respectively positive ionized acceptors and donors.

$$\int \frac{d\mathbf{p}}{(2\pi\hbar)^3} (f_e(\mathbf{p}) - f_h(\mathbf{p})) + nf_-^A = 0$$

in the p-material and

$$\int \frac{d\mathbf{p}}{(2\pi\hbar)^3} (f_e(\mathbf{p}) - f_h(\mathbf{p})) - n_D f_+^D = 0$$

in the n-material. (Here f_e, f_h and f_-^A, f_+^D are the Fermi distributions of electrons and holes in the bands, respectively on the donors and acceptors.) The chemical potentials in these two equations μ_p and μ_n are of course different and $\mu_p < \mu_n$.

Now let us consider the simplest case of zero temperature (see Fig. 2.23). In the ground state of the p-type semiconductor the valence band (edge - red line) is full occupied, while the acceptor level (dotted blue line) and the conduction band (edge - blue line) are empty, the chemical potential (green line) lies somewhere between the valence band and the acceptor level. In the ground state of the n-type semiconductor

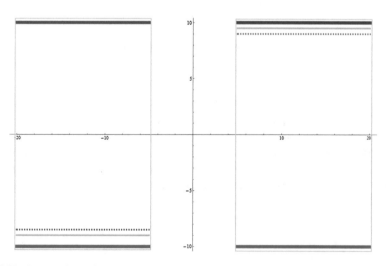

Fig. 2.23 Separated p and n semiconductors at $T = 0$

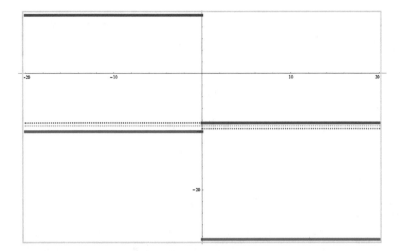

Fig. 2.24 p and n semiconductors in contact at $T = 0$

the valence band (red) and the donor level (dotted red line) are occupied and the chemical potential (green line) lies between the donor level and the conduction band.

When the contact between the semiconductors is established in thermal equilibrium at $T = 0$ one has a single system where the chemical potential μ must be the same overall (green line). This is possible only if at the contact surface a potential jump occurs as it is shown on Fig. 2.24.

Assuming the contact is defined by a plane transverse to the x -axis at $x = 0$, at any finite temperature T one has an in-homogeneous charge density of electrons and holes:

$$\rho(x) = e \int \frac{d\mathbf{p}}{(2\pi\hbar)^3} \cdot (f_e(\mathbf{p},x) - f_h(\mathbf{p},x)) - en_D\theta(x)f_+^D(x) + en_A\theta(-x)f_-^A(x)$$

with

$$f_e(\mathbf{p}, x) = \frac{1}{e^{\beta\left(\frac{p^2}{2m_e}+eV(x)+\frac{1}{2}Eg-\mu\right)} + 1}$$

$$f_h(\mathbf{p}, x) = \frac{1}{e^{\beta\left(\frac{p^2}{2m_h}-eV(x)+\frac{1}{2}Eg+\mu\right)} + 1}$$

and

$$f_-^A(x) = \frac{1}{e^{\beta(E_A+eV(x)-\mu)} + 1}$$

$$f_+^D(x) = 1 - \frac{1}{e^{\beta(E_D+eV(x)-\mu)} + 1} = \frac{1}{e^{\beta(-E_D-eV(x)+\mu)} + 1} \, .$$

Here the reference energy choice was the middle of the gap and therefore E_A and E_D are the distances to the center of the gap.

The potential $V(x)$ has to be determined self-consistently from the Poisson equation

$$\varepsilon \frac{d^2}{dx^2} V(x) = -4\pi \rho(x)$$

with the condition of continuity of the derivative of the potential at $x = \pm 0$:

$$\frac{d}{dx} V(x)|_{x=-0} = \frac{d}{dx} V(x)|_{x=+0}$$

(since the dielectric constants of the two materials were assumed to be the same). The potential energy has to be assumed also to be continuous and therefore $V(-0) = V(+0)$. To solve the Poisson equation (a second order differential equation) one still needs two initial conditions. Usually one gives the initial value of the function and its derivative in the initial point. Here the peculiarity of the physics defines two different conditions to impose, namely the values at $\pm\infty$, that should correspond to the situation in the separated semiconductors. These two conditions define uniquely the solution, however these equations are not solvable analytically and the numerical method to solve it must be correspondingly adapted.

We have seen before the solution for $T = 0$. For higher temperatures (in the physically relevant situations this is just the room temperature) one may consider the non-degenerate case, that simplifies essentially the equations.

Indeed in the non-degenerate case the charge density is given by

$$\rho(x) = e \int \frac{d\mathbf{p}}{(2\pi\hbar)^3} \left(e^{-\beta\left(\frac{p^2}{2m_e} + eV(x) + \frac{1}{2}Eg - \mu\right)} - e^{-\beta\left(\frac{p^2}{2m_h} - eV(x) + \frac{1}{2}Eg + \mu\right)} \right)$$
$$+ \theta(x) e n_D e^{\beta(E_D + eV(x) - \mu)} - \theta(-x) e n_A e^{-\beta(E_A + eV(x) - \mu)} .$$

The integrals over the momenta may be performed and we get

$$\rho(x) = e n_{0e} e^{-\beta(eV(x) - \mu)} - e n_{0h} e^{\beta(eV(x) - \mu)}$$
$$- \theta(x) e n_D^0 e^{\beta(eV(x) - \mu)} + \theta(-x) e n_A^0 e^{-\beta(eV(x) - \mu)}$$

where $n_{0e} = \left(\frac{2\pi\hbar^2}{k_B T m_e}\right)^{\frac{3}{2}} e^{-\beta\frac{1}{2}E_g}$, $n_{0h} = \left(\frac{2\pi\hbar^2}{k_B T m_e}\right)^{\frac{3}{2}} e^{-\beta\frac{1}{2}E_g}$ and $n_D^0 = n_D e^{\beta E_D}$, $n_A^0 = n_A e^{-\beta E_A}$. (Actually, only the non-degeneracy of the distribution in the bands is essential for this step!)

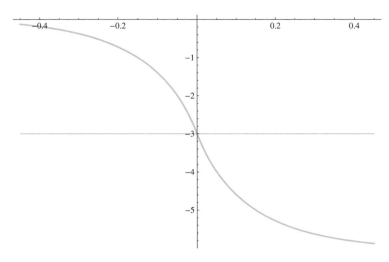

Fig. 2.25 $\beta eV(x)$ in the non-degenerate case (with $c = 3$)

We may require a total charge neutrality on both sides (no macroscopic polarization!). Taking into account that the potential differs from constants $V(\pm\infty)$ only in a finite domain in the vicinity of the contact, we may write

$$e\left(n_{0e}e^{-\beta(eV(-\infty)-\mu)} - en_{0h}e^{\beta(eV(-\infty)-\mu)} + n_A^0 e^{-\beta(eV(-\infty)-\mu)}\right) = 0$$
$$e\left(n_{0e}e^{-\beta(eV(+\infty)-\mu)} - en_{h}e^{\beta(eV(+\infty)-\mu)} - n_D^0 e^{-\beta(eV(+\infty)-\mu)}\right) = 0$$

So long the semiconductors were separated both equations were satisfied with $V(\pm\infty) = 0$ and the chemical potentials μ_p, respectively μ_n of the two materials respectively.

After the contact has been established one may still choose $V(-\infty) = 0$ and therefore, since far away from the contact the position of the chemical potential relative to the bands has to be the same as before, one must take $\mu \equiv \mu_n$. On the other hand, the same must be true at $x \to \infty$ for the other semiconductor. It follows, that simultaneously $\mu \equiv \mu_n + eV(\infty)$. Therefore $eV(\infty) = \mu_p - \mu_n$. Then, not only the electro-neutrality is overall satisfied, but the charge density vanishes at $x \to \pm\infty$ and all the parameters are well defined.

The non-degenerate solution, obtained numerically for the simplified symmetrical case of $n_A = n_D$ shows, on a convenient length scale a smoothing of the potential drop at the contact as in Fig. 2.25 with the choice $c \equiv \frac{1}{2}\beta(\mu_n - \mu_p) = 3$.

To conclude, the p-n contact in equilibrium creates a potential barrier for the carriers. An application of an external electric potential either lowers or increases this barrier and therefore the current flow depends on the applied polarity. The p-n contact acts as a rectifier.

Chapter 3
Electron-Electron Interaction

The potential a valence electron feels in a crystal is not a predetermined one. The electrons interact with each other by Coulomb forces and the one-electron theory is to be understood at most in the sense of a self-consistent approximation. Diagrammatic perturbation theory may lead to higher many-body corrections. In a semiconductor an electron in the conduction band may move "freely", while in the valence band the same is true for the holes (absent electrons). This leads to the picture of an electron-hole plasma. A single electron-hole pair by Coulomb attraction may be bound to an exciton with hydrogen-like spectrum.

3.1   The Exciton

3.1.1   Wannier Exciton

Until now we discussed the electron states and their occupation in a periodical lattice under the assumption of no interaction between the electrons. We nevertheless made an exception to correct the double occupation of the localized states. In this chapter, we shall discuss other many-electron aspects. The next simplest problem is the Coulomb attraction between a conduction band electron and a valence band hole in a semiconductor. Obviously these two may form a bound state, called exciton. As in the description of the impurity states we use the effective mass approximation in the neighborhood of the band extrema. By separating the center of mass motion

$$\Psi(\mathbf{r}_e, \mathbf{r}_h) = \phi(\mathbf{r}_{CM})\chi(\mathbf{r}_{rel})$$

© Springer International Publishing AG, part of Springer Nature 2018
L. A. Bányai, *A Compendium of Solid State Theory*,
https://doi.org/10.1007/978-3-319-78613-1_3

$$\mathbf{r}_{CM} \equiv \frac{m_e^* \mathbf{r}_e + m_h^* \mathbf{r}_h}{m_e^* + m_h^*}; \qquad \mathbf{r}_{rel} \equiv \mathbf{r}_e - \mathbf{r}_h$$

$$\phi(\mathbf{r}_{CM}) = \frac{1}{\sqrt{V}} e^{i\mathbf{k}_{CM}\mathbf{r}_{CM}} \quad ,$$

the relative motion is described by the Wannier equation

$$\left\{ -\frac{\hbar^2}{2\mu}\nabla^2 - \frac{e^2}{\varepsilon|\mathbf{r}|} + E_g - E \right\} \chi(\mathbf{r}) = 0$$

where the reduced mass is given by $\frac{1}{\mu} = \frac{1}{m_e^*} + \frac{1}{m_h^*}$. As it is well known from the hydrogen atom problem, the eigenstates are characterized by the quantum numbers $n = 1, 2, \ldots, l = 0, 1 \ldots, m = 0 \pm 1, \ldots$, while the eigenenergies depend only on the main quantum number n. The ground state wave function is

$$\psi_{1,0} = \frac{1}{\sqrt{\pi} a_B^{\frac{3}{2}}} e^{-\frac{r}{a_B}}$$

with the ground state energy

$$E_n^x = E_g - E_R \,,$$

where the Bohr radius and the Rydberg energy are

$$a_B = \frac{\hbar^2 \varepsilon}{e^2 \mu}, \quad E_R = \frac{e^2}{2\varepsilon a_B} = \frac{\hbar^4}{2\mu a_B^2} \,.$$

However, as we already discussed by the ionic impurity states, the scales of energy and radius are essentially different from that of the hydrogen atom due to the difference in the masses and the dielectric constant.

This state may not be represented in the one particle state density (it does not belong to the one-electron states), but may be seen in the absorption spectrum. For higher lying states of the pair the above described approximation is however not appropriate.

3.1.2 Exciton Beyond the Effective Mass Approximation

We shall describe here an alternative way to treat the exciton states without using the effective mass approximation. The Hamiltonian of the conduction band electrons and valence band holes of spin σ in the second quantization formalism may be defined either through the creation/annihilation operators of the Bloch or of the Wannier states

$$H_0 = \sum_{\sigma=\pm 1} \sum_{\alpha=e,h} \sum_{k \in BZ} \varepsilon_\alpha(\mathbf{k}) a^+_{\alpha,\sigma,\mathbf{k}} a_{\alpha,\sigma,\mathbf{k}}$$

$$= \sum_{\sigma=\pm 1} \sum_{\alpha=e,h} \sum_{\mathbf{R},\mathbf{R}'} t_\alpha(\mathbf{R} - \mathbf{R}') a^+_{\alpha,\sigma,\mathbf{R}} a_{\alpha,\sigma,\mathbf{R}'} ; \quad (\alpha = e, h) .$$

Here $\varepsilon_\alpha(\mathbf{k})$ $(\alpha = e, h)$ are the band energies of electrons respectively, holes. The Hamiltonian of the Coulomb attraction between the electrons and holes, while being defined through the charge densities, is easily expressed in terms of the second quantized wave functions

$$H_{eh} = -\sum_{\sigma,\sigma'} \int dx \int dx' \psi_{e,\sigma}(\mathbf{x})^+ \psi_{e,\sigma}(\mathbf{x}) \frac{e^2}{|\mathbf{x} - \mathbf{x}'|} \psi^+_{h,\sigma'}(\mathbf{x}') \psi_{h,\sigma'}(\mathbf{x}') .$$

The second quantized wave functions of conduction band electrons or valence band holes of spin σ may be again defined through the annihilation/creation operators either of Bloch states or Wannier states

$$\psi_{\alpha,\sigma}(\mathbf{x}) = \sum_{\mathbf{k}} e^{i\mathbf{k}\mathbf{x}} u_{\alpha\mathbf{k}}(\mathbf{x}) a_{\alpha,\sigma,\mathbf{k}} = \sum_{\mathbf{R}} w_\alpha(\mathbf{x}, \mathbf{R}) a_{\alpha,\sigma,\mathbf{R}} ; \quad (\alpha = e, h).$$

In the Wannier version one may use the fact, that the Wannier functions are strongly localized on the lattice nodes in order to introduce a simplifying approximation. Namely, to retain in the charge density operator only the terms corresponding to the local densities on the nodes i.e.

$$H_{eh} \approx -\sum_{\sigma,\sigma'} \sum_{\mathbf{R},\mathbf{R}'} \int dx \int dx' |w_e(\mathbf{x}, \mathbf{R})|^2 \frac{e^2}{|\mathbf{x} - \mathbf{x}'|} |w_h(\mathbf{x}, \mathbf{R}')|^2 a^+_{e,\sigma,\mathbf{R}} a_{e,\sigma,\mathbf{R}} a^+_{h,\sigma,\mathbf{R}'} a_{h,\sigma,\mathbf{R}'} .$$

Furthermore, since the Coulomb potential varies slowly on the scale of the elementary cell we may even write

$$H_{eh} \approx -\sum_{\sigma=\pm 1} \sum_{\sigma'=\pm 1} \sum_{\mathbf{R} \neq \mathbf{R}'} \frac{e^2}{|\mathbf{R} - \mathbf{R}'|} a^+_{e,\sigma,\mathbf{R}} a_{e,\sigma,\mathbf{R}'} a^+_{h,\sigma',\mathbf{R}'} a_{h,\sigma',\mathbf{R}}$$

$$- \sum_{\sigma=\pm 1} \sum_{\sigma'=\pm 1} c_{eh} \sum_{\mathbf{R}} a^+_{e,\sigma,\mathbf{R}} a_{e,\sigma,\mathbf{R}} a^+_{h,\sigma',\mathbf{R}} a_{h,\sigma',\mathbf{R}} .$$

Here we have separated the on-site terms $\mathbf{R} = \mathbf{R}'$ since in that case one has still to compute the finite integral

$$c_{eh} = \int dx \int dx' |w_e(\mathbf{x}, \mathbf{R})|^2 \frac{e^2}{|\mathbf{x} - \mathbf{x}'|} |w_h(\mathbf{x}, \mathbf{R})|^2 .$$

One may encounter two extreme cases:

(i) either the on-site term c_{eh} dominates, and one may ignore the inter-site terms (Frenkel exciton) with the lattice Hamiltonian

$$H_{Frenkel} = \sum_{\sigma=\pm 1} \sum_{\alpha=e,h} \sum_{\mathbf{R},\mathbf{R}'} t_\alpha (\mathbf{R} - \mathbf{R}') a^+_{\alpha,\sigma,\mathbf{R}} a_{\alpha,\sigma,\mathbf{R}'}$$
$$- \sum_{\sigma=\pm 1} \sum_{\sigma'=\pm 1} c_{eh} \sum_{\mathbf{R}} a^+_{e,\sigma,\mathbf{R}} a_{e,\sigma,\mathbf{R}} a^+_{h,\sigma',\mathbf{R}} a_{h,\sigma',\mathbf{R}},$$

(ii) or the inter-site term dominates and we get the Wannier Hamiltonian

$$H_{Wannier} = \sum_{\sigma=\pm 1} \sum_{\alpha=e,h} \sum_{\mathbf{R},\mathbf{R}'} t_\alpha (\mathbf{R} - \mathbf{R}') a^+_{\alpha,\sigma,\mathbf{R}} a_{\alpha,\sigma,\mathbf{R}'}$$
$$- \sum_{\sigma=\pm 1} \sum_{\sigma'=\pm 1} \sum_{\mathbf{R}\neq\mathbf{R}'} \frac{e^2}{|\mathbf{R} - \mathbf{R}'|} a^+_{e,\sigma,\mathbf{R}} a_{e,\sigma,\mathbf{R}'} a^+_{h,\sigma',\mathbf{R}'} a_{h,\sigma',\mathbf{R}}$$

Accordingly, in the Wannier case we get for the energy of a single electron hole pair the Hamiltonian matrix

$$h_{\mathbf{R},\mathbf{R}'} = t_e (\mathbf{R} - \mathbf{R}') + t_h (\mathbf{R} - \mathbf{R}') - \frac{e^2}{|\mathbf{R} - \mathbf{R}'|} (1 - \delta_{\mathbf{R},\mathbf{R}'})$$

or after a discrete Fourier transformation (implying that the wave vectors belong to the Brillouin Zone)

$$h_{\mathbf{k},\mathbf{k}'} = (\varepsilon_e(\mathbf{k}) + \varepsilon_h(\mathbf{k})) \delta_{\mathbf{k},\mathbf{k}'} - V(\mathbf{k} - \mathbf{k}')$$

with

$$V(\mathbf{q}) = \sum_{\mathbf{R}\neq 0} \frac{e^2}{|\mathbf{R}|} e^{i\mathbf{q}\mathbf{R}} .$$

What we have gained is that now we included all the band states and therefore, we may compute also, higher lying pair states influenced by the Coulomb attraction. This is illustrated for the Wannier case in Fig. 3.1 showing the calculated exciton state density (electron-hole pair states) with and without Coulomb interaction in a cubic crystal, within the simplified tight binding model with nearest neighbor coupling (see Sect. 2.3.7). In the absence of Coulomb forces, one may remark in the middle of the spectrum the sharp flat top (Van Hove singularity), which is a consequence of the discontinuous jumps of the Brillouin zone.

Fig. 3.1 Exciton state density in the whole band calculated within the Wannier model. (black—without Coulomb interaction, blue-with Coulomb interaction)

3.2 Many-Body Approach to the Solid State

3.2.1 Self-consistent Approximations

Until now we have tried to describe the properties of crystals only within the frame of one particle (or at most two-) particle states in the presence of a periodical potential. This potential however may not be identified with the potential created by the lattice of the bare ions, since the electrons at their turn screen the ions. Each electron sees not only the ions, but also the other electrons. To apply the one-particle treatment we need a way to define the periodical potential to be used for the calculation of the band structure.

The most important schemes for the treatment of quantum mechanical many-body systems are the self-consistent approximations we describe here. Within this frame it is possible also to precise the definition of the periodic potential. (In what follows, for the simplicity of notations, we include the coordinate $\mathbf{r}$ as well as the spin $\sigma = \pm\frac{1}{2}$ in the generalized coordinate x, whenever it is meaningful.)

We consider here a system of electrons interacting through repulsive Coulomb forces in the presence of an external potential $U(x)$. The later one includes the attractive Coulomb potential of the positive ions, supposed to stay on a periodical lattice. The dynamics of the ions will be treated later in the chapter about phonons. Within the second quantized formalism (see Chap. 9) the Hamilton operator of this system is

$$H = \int dx \psi(x)^+ \left[-\frac{\hbar^2}{2m} \nabla^2 + U(x) \right] \psi(x)$$
$$+ \frac{1}{2} \int dx \int dx' \psi(x)^+ \psi(x')^+ \frac{e^2}{|x - x'|} \psi(x')\psi(x) .$$

The purpose is to reduce the problem in a first approximation to an effective one-particle one (consisting only of products of two creation-annihilation operators) in order to be tractable and close to the previous discussion of the quantum-mechanical motion one-particle in a periodical potential.

The simplest scheme consists in "developing" the product of two operators A and B around their averages. Let us consider the identity

$$AB = A\langle B \rangle + \langle A \rangle B - \langle A \rangle \langle B \rangle + (A - \langle A \rangle)(B - \langle B \rangle) .$$

Or, ignoring the "fluctuation" around the average values

$$(A - \langle A \rangle)(B - \langle B \rangle)$$

we have

$$AB \approx A\langle B \rangle + \langle A \rangle B - \langle A \rangle \langle B \rangle$$

Using this approximation for the choice $A \equiv \psi(x)^+ \psi(x)$ and $B \equiv \psi(x')^+ \psi(x')$ in the interaction term we get the one-electron type Hartree Hamiltonian

$$H^{\mathscr{H}} \equiv \int dx \psi(x) \left\{ -\frac{\hbar^2}{2m} \nabla^2 + U_{eff}^H(x) \right\} \psi(x) + const^{\mathscr{H}}$$

with the effective Hartree potential

$$U_{eff}^{\mathscr{H}}(x) \equiv U(x) + \sum_{\sigma'} \int dx' \frac{e^2}{|x - x'|} \langle \psi_{\sigma'}(x')^+ \psi_{\sigma'}(x') \rangle$$

and the Hartree constant

$$const^{\mathscr{H}} = -\frac{1}{2} \int dx \int dx' \frac{e^2}{|x - x'|} \langle \psi_\sigma(x)^+ \psi_\sigma(x) \rangle \langle \psi_{\sigma'}(x')^+ \psi_{\sigma'}(x') \rangle .$$

This constant ensures that $\langle H^{\mathscr{H}} \rangle = \langle H \rangle$.

Until now we did not specify the meaning of the averages themselves. They may be defined by their equilibrium values i.e.

$$\langle \psi(x)^+ \psi(x) \rangle_0 \equiv \frac{1}{Tr \left\{ e^{-\beta(H^{\mathscr{H}} -\mu N)} \right\}} Tr \left\{ e^{-\beta(H^{\mathscr{H}} -\mu N)} \psi(x)^+ \psi(x) \right\}$$

and computed from this definition self-consistently. The result is a temperature dependent effective one-particle spectrum with Hartree eigenfunctions and eigenvalues

$$\left\{-\frac{\hbar^2}{2m}\nabla^2 + U_{eff}^{\mathcal{H}}(x)\right\}\phi(x) = \varepsilon\phi(x) \ .$$

The resulting effective Bloch spectrum in the presence of only a periodical potential due to the fixed ions is therefore relevant for a given temperature. The potential to be used in calculations of the Bloch functions and band spectrum in a real crystal is this self-consistent one. One uses mostly the zero temperature (ground state) as reference state.

However, on may tackle also a time dependent problem within a s.c. approximation, with time dependent averages $\langle\psi(x)^+\psi(x)\rangle_t$ with initial conditions and additional non-periodical external potentials. In this case the Hartree Hamiltonian is implicitly time dependent. Nevertheless, if the external potential does not depend on time, the average energy is conserved (again due to the special role of the time-dependent Hartree constant!)

$$\frac{\partial}{\partial t}\langle H^{\mathcal{H}}(t)\rangle_t = 0 \ .$$

The choice of the operators A and B in our former reasoning was arbitrary. The proper argument is, that as it may be shown, the Hartree choice gives the best one-particle approximation for the grand canonical distribution conserving the total number of fermions and being diagonal in the coordinates and spin.

An ever better one particle approximation for the grand canonical distribution is the Hartree-Fock one, allowing also for terms non-diagonal in the coordinates and spins. The former simple arguing however cannot provide for it, The Hartree-Fock approximation implies a non-diagonal effective potential

$$H^{\mathcal{H}F} \equiv -\int dx\psi(x)\frac{\hbar^2}{2m}\nabla^2\psi_\sigma(x) + \int dx \int dx'\psi(x)^+ U_{eff}^{\mathcal{H}F}(x,x')\psi(x') + const^{\mathcal{H}F}$$

defined as

$$U_{eff}^{\mathcal{H}F}(x,x') \equiv \delta(x-x')\int dy \left(\frac{e^2}{|x-y|}\langle\psi(y)^+\psi(y)\rangle\right) - \frac{e^2}{|x-x'|}\langle\psi(x')^+\psi(x)\rangle$$

with the Hartree Fock constant

$$const^{\mathcal{H}F} \equiv \frac{1}{2}\int dx \int dx'\frac{e^2}{|x-x'|} \times$$
$$\left[\langle\psi(x)^+\psi(x)\rangle\langle\psi(x')^+\psi(x')\rangle - \langle\psi(x)^+\psi(x')\rangle\langle\psi(x')^+\psi(x)\rangle\right]$$

Again, it may be shown, that $\langle H^{\mathscr{H}F} \rangle = \langle H \rangle$ and its time-dependent variant conserves the average energy in the absence of time dependent external potential. The new Coulomb term containing the non-local average $\langle \psi(x')^+ \psi(x) \rangle$ is called Coulomb exchange energy.

In both schemes, one has reduced the many-body problem to a self-consistent one particle problem. However, for the treatment of some phase transitions with spontaneous symmetry breaking of the particle number conservation (like in the case of Bose condensation or superconductivity we shall discuss in Chap. 7) one needs generalizations of the Hartree-Fock scheme in keeping also particle number non-conserving averages as $\langle \psi(x')\psi(x) \rangle$.

3.2.2   Electron Gas with Coulomb Interactions

As a simple illustration of the Hartree-Fock approximation we consider here a gas of electrons that interact through Coulomb forces, without any periodical potential at zero temperature (ground state). To stabilize the system without losing translational invariance one needs however to consider a uniform positive background to replace the positive ions. In the absence of a magnetic field we have spin independence and therefore,

$$\langle \psi_\sigma(\mathbf{r})^+ \psi_{\sigma'}(\mathbf{r}') \rangle = \delta_{\sigma,\sigma'} \eta(\mathbf{r}\,\mathbf{r}')$$

while from the homogeneity it follows

$$\eta(\mathbf{r}, \mathbf{r}') \equiv \eta(|\mathbf{r} - \mathbf{r}'|).$$

Then the Hartree-Fock Hamiltonian is

$$H^{\mathscr{H}F} = \sum_{\sigma,\sigma'} \int d\mathbf{r} \int d\mathbf{r}' \psi_\sigma(\mathbf{r}) h^{\mathscr{H}F}_{\sigma\sigma'}(\mathbf{r}, \mathbf{r}') \psi_{\sigma'}(\mathbf{r}')$$

$$h^{\mathscr{H}F}_{\sigma\sigma'}(\mathbf{r}, \mathbf{r}') = \delta_{\sigma\sigma'} \left\{ h_0(\mathbf{r}, \mathbf{r}') + \delta(\mathbf{r} - \mathbf{r}') \int d\mathbf{r}'' \frac{\langle n \rangle e^2}{|\mathbf{r} - \mathbf{r}''|} - \frac{e^2}{|\mathbf{r} - \mathbf{r}'|} \eta(\mathbf{r} - \mathbf{r}') \right\}$$

with

$$h_0(\mathbf{r}, \mathbf{r}') \equiv \delta(\mathbf{r} - \mathbf{r}') \left(-\frac{\hbar^2}{2m} \nabla^2 + V_+ \right).$$

The term

$$V_+ = -\langle n \rangle e^2 \int d\mathbf{r}' \frac{1}{|\mathbf{r}'|}$$

represents the Coulomb potential energy due to the uniform positive background charge and $\langle n \rangle$ is the average density of electrons ($\langle n \rangle = 2\eta(0)$).

Of course, all the homogeneity assumptions are true only in an infinite system and this constant potential energy V_+ is divergent. However, we keep for a wile the volume finite and as we shall see, this diverging entity will compensate other inherently diverging entities. Without the positive background contribution, the positive energy the Coulomb repulsion between the electrons would push them far away, without reaching a stable state.

Due to the homogeneity, the Hartree-Fock eigenfunctions have to be plane waves

$$\phi_{\mathbf{k}}(\mathbf{r}) = \frac{1}{\sqrt{\Omega}} e^{-\imath \mathbf{kr}}$$

and therefore with the step-like Fermi function $\theta(k_F - |\mathbf{k}|)$ at $T = 0$ the average density is

$$\langle \psi_\sigma(\mathbf{r})^+ \psi_{\sigma'}(\mathbf{r}') \rangle \equiv \delta_{\sigma,\sigma'} \frac{1}{\Omega} \sum_{|\mathbf{k}|<k_F} e^{\imath \mathbf{k}(\mathbf{r}-\mathbf{r}')} \ .$$

Then the Hartree-Fock eigenvalue equation

$$\left\{ -\frac{\hbar^2}{2m}\nabla^2 - V_+ + \frac{1}{\Omega} \sum_{|\mathbf{k}'|<k_F} \int d\mathbf{r}' \frac{e^2}{|\mathbf{r}-\mathbf{r}'|} \left[2 - e^{\imath(\mathbf{k}-\mathbf{k}')(\mathbf{r}-\mathbf{r}')} \right] \right\} e^{-\imath \mathbf{kr}} = E_k e^{-\imath \mathbf{kr}}$$

just defines the energy eigenvalues E_k as

$$E_k = \frac{\hbar^2 k^2}{2m} - \frac{e^2}{\Omega} \sum_{|\mathbf{k}'|<k_F} \int d\mathbf{r}' e^{\imath(\mathbf{k}'-\mathbf{k})(\mathbf{r}-\mathbf{r}')} \frac{1}{|\mathbf{r}-\mathbf{r}'|} \ .$$

Using the (improper) definition of the Fourier transform of the Coulomb potential by the limit procedure

$$\lim_{\kappa \to 0} \int d\mathbf{r} \frac{e^2 e^{-\kappa r}}{r} e^{\imath \mathbf{qr}} = \frac{4\pi e^2}{q^2}$$

one gets finally

$$E_k = \frac{\hbar^2 k^2}{2m} - \frac{e^2 k_F}{4\pi} \left(2 + \frac{k_F^2 - k^2}{kk_F} \ln \left| \frac{k + k_F}{k - k_F} \right| \right) \ .$$

The total average energy of an electron of the Coulomb electron gas (per spin) differs from that of the free electron gas in the ground state $\frac{3}{5}E_F$

$$\langle E \rangle \equiv \frac{\sum_{|\mathbf{k}|<k_F} E_k}{\sum_{|\mathbf{k}|<k_F} 1} = \frac{3}{5}E_F - \frac{3e^2 k_F}{8\pi}$$

by the "exchange energy" $-\frac{3e^2 k_F}{8\pi}$.

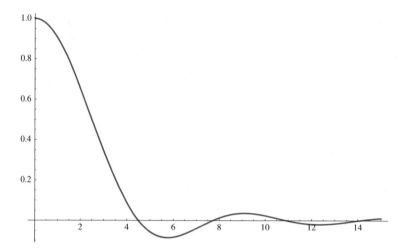

Fig. 3.2 Exchange correlation function

The "exchange correlation function" given by

$$\frac{1}{n}\eta(\mathbf{r}) = \frac{1}{n}\sum_{\sigma} < \psi_{\sigma}(0)^{+}\psi_{\sigma'}(\mathbf{r}) > = \frac{1}{\Omega}\sum_{|\mathbf{k}|<k_F} e^{-i\mathbf{k}\mathbf{r}} = -\frac{3}{2(k_F r)^3}(k_F r \cos k_F r - \sin k_F r)$$

defines the probability amplitude to find an electron at the distance $r \equiv |\mathbf{r}|$ as function of the variable $k_F r$ and it looks as in Fig. 3.2.

3.2.3   The Electron-Hole Plasma

While the previously described model is used with reference to metals, the model of a Coulomb interacting electron-hole plasma we describe here tries to catch the main many-body aspects in a semiconductor. We have seen, that unoccupied electron states in a valence band with negative effective mass act as positively charged particles (holes) with positive effective mass. The negatively charged electrons in the conduction band (from now on just called—electrons, if not otherwise specified) and the positively charged holes interact by Coulomb forces. The Hamiltonian of such a plasma in terms of the second quantized wave functions of elections and holes

$$\psi_{e,\sigma}(\mathbf{r}) = \sum_{\mathbf{k}} \frac{1}{\sqrt{\Omega}} e^{i\mathbf{k}\mathbf{r}} a_{ek\sigma} \quad ; \quad \psi_{h,\sigma}(\mathbf{r}) = \sum_{\mathbf{k}} \frac{1}{\sqrt{\Omega}} e^{i\mathbf{k}\mathbf{r}} a_{hk\sigma}$$

is given by

$$
H = \sum_{\mathbf{k},\sigma} (\frac{\hbar^2}{2m_e}k^2 + \frac{1}{2}E_g)a^+_{e,\mathbf{k},\sigma}a_{e,\mathbf{k},\sigma} + \sum_{\mathbf{k},\sigma} (\frac{\hbar^2}{2m_h}k^2 + \frac{1}{2}E_g)a^+_{h,\mathbf{k},\sigma}a_{h,\mathbf{k},\sigma} + E_0
$$

$$
+ \frac{1}{2}\sum_{\sigma,\sigma'} \int d\mathbf{r} \int d\mathbf{r}' \frac{e^2}{|\mathbf{r}-\mathbf{r}'|} \{\psi_{e,\sigma}(\mathbf{r})^+\psi_{e\sigma'}(\mathbf{r}')^+\psi_{e\sigma'}(\mathbf{r}')\psi_{e\sigma}(\mathbf{r})
$$

$$
+ \psi_{h,\sigma}(\mathbf{r})^+\psi_{h\sigma'}(\mathbf{r}')^+\psi_{h\sigma'}(\mathbf{r}')\psi_{h\sigma}(\mathbf{r})
$$

$$
- 2\psi_{e,\sigma}(\mathbf{r})^+\psi_{e\sigma}(\mathbf{r})\psi^+_{h\sigma'}(\mathbf{r}')\psi_{h\sigma'}(\mathbf{r}')\} .
$$

Here one ignores the Bloch character of the wave functions. The electrons and holes are considered just as free propagating particles. This Hamiltonian conserves the number of electrons N_e as well as of the holes N_h, while the original Coulomb interaction of the conduction and valence band electrons

$$
\frac{1}{2}\int dx \int dx' \psi(\mathbf{r})^+\psi(\mathbf{r}')^+ \frac{e^2}{|\mathbf{r}-\mathbf{r}'|}\psi(\mathbf{r}')\psi(\mathbf{r})
$$

conserves the total number of electrons in both bands, or otherwise stated it only conserves the difference between the number of electrons and holes $N_e - N_h$. Indeed, the second quantized wave function of conduction and valence band electrons is by definition $\psi_\sigma(\mathbf{x}) = \psi_{e,\sigma}(\mathbf{x}) + \psi_{h,-\sigma}(\mathbf{x})^+$ and therefore, the Coulomb interaction term contains not only conduction band electron and valence band hole charge densities, but also mixed ("exchange") terms.

However, since the Coulomb potential varies slowly within the elementary cell, these terms are small. The matrix element of a function $F(\mathbf{r})$ between Bloch states may be written as a sum of integrals over the elementary cell v

$$
\int d\mathbf{r} u^\alpha_{\mathbf{k}}(\mathbf{r})^* u^\beta_{\mathbf{k}'}(\mathbf{r})F(\mathbf{r}) = \sum_{\mathbf{R}} \int_v d\mathbf{r} u^\alpha_{\mathbf{k}}(\mathbf{r})^* u^\beta_{\mathbf{k}'}(\mathbf{r})F(\mathbf{r})
$$

If the function $F(\mathbf{r})$ varies slowly within the elementary cell, then we may replace it by its value in the cell (v) i.e. we get

$$
\approx \sum_{\mathbf{R}} F(\mathbf{R}) \int_v d\mathbf{r} u^\alpha_{\mathbf{k}}(\mathbf{r})^* u^\beta_{\mathbf{k}'}(\mathbf{r})
$$

and further at small wave vectors

$$
\approx \sum_{\mathbf{R}} F(\mathbf{R}) \int_v d\mathbf{r} u^\alpha_0(\mathbf{r})^* u^\beta_0(\mathbf{r}) = \delta_{\alpha,\beta}\sum_{\mathbf{R}} F(\mathbf{R}) .
$$

In the electron-hole plasma model ignored "exchange" terms contain such integrals with $\alpha \neq \beta$ and therefore may be neglected.

Nevertheless, as we shall see later, the interaction with electromagnetic fields (photons) may create or annihilate electron-hole pairs and this is the basic process in the optics of semiconductors.

3.2.4   Many-Body Perturbation Theory of Solid State

Beyond the Hartree-Fock approximation one has at disposal just the diagrammatic perturbation theory to improve it. Of course, within this frame one has developed sophisticated methods, but the first question is how to start. In solid state theory, it seems not too clever to consider the motion of the electrons in the field of the rigid ions as the unperturbed problem and to treat the Coulomb interaction as a perturbation. A better way is to start from the Hartree-Fock approximation and treat the difference to the true Hamiltonian as perturbation. According Sect. 3.2.1

$$
\begin{aligned}
H' = H - H^{\mathscr{H}F} = \frac{1}{2}\int dx \int dx' \frac{e^2}{|x-x'|} & \{\psi(x)^+\psi(x')^+\psi(x')\psi(x) \\
& - 2\langle\psi^+(x')\psi(x')\rangle\psi^+(x)\psi(x) + 2\langle\psi^+(x')\psi(x)\rangle\psi^+(x)\psi(x') \\
& + \langle\psi^+(x)\psi(x)\rangle\langle\psi^+(x')\psi(x')\rangle - \langle\psi^+(x')\psi(x)\rangle\langle\psi^+(x)\psi(x')\rangle\}
\end{aligned}
$$

In terms of the Hartree-Fock one-electron eigenstates $\phi_\nu(x)$ the second quantized wave function of the electron in the crystal is

$$
\psi(x) = \sum_\nu A_\nu\phi_\nu(x)
$$

with A_ν being the annihilation operator of the electron state with quantum number ν. This quantum number may be the band index and wave vector of a Bloch state, but we do not need to specify it here. In terms of these operators the Hartree-Fock Hamiltonian is

$$
H^{\mathscr{H}F} = \sum_\sigma\sum_\nu \varepsilon_\nu A_{\nu,\sigma}^+ A_{\nu,\sigma} + const.^{\mathscr{H}F}
$$

We choose the Hartree-Fock ground state ($T = 0$) for the averages. also all the states with energy below the Fermi energy $\varepsilon_\nu < \varepsilon_F$ being occupied, while above it are all empty. The averages themselves are then

$$
\langle\psi^+(x)\psi(x')\rangle = \sum_\nu \phi_\nu^*(x)\phi_\nu(x')\theta(\varepsilon_F - \varepsilon_\nu)
$$

Now it is useful to introduce new electron-hole annihilation (and creation) operators $a_{e,\nu}$ (for $\varepsilon_\nu > \varepsilon_F$) and $a_{h,\nu}$ (for $\varepsilon_\nu < \varepsilon_F$) by

$$A_v = \begin{cases} a_{e,v} & (\varepsilon_v > \varepsilon_F) \\ a_{h,\tilde{v}}^+ & (\varepsilon_v < \varepsilon_F) \end{cases}$$

For the Bloch indices, if $v = n$, $\mathbf{k}$, σ the conjugate quantum number is $\tilde{v} = n$, $-\mathbf{k}$, $-\sigma$. Within the new description the many-body ground state $|0>$ is the electron-hole vacuum

$$a_{e,v}|0>= 0; \quad a_{h,v}|0>= 0 .$$

If one expresses the perturbation H' in terms of the electron-hole creation-annihilation operators and performs all the necessary commutations such, that all the electron-hole annihilation operators a_v are on the right side, while the creation operators a_v^+ are on the left side, then one may see that all the supplementary terms disappear. In other words, the Coulomb terms are equivalent to the "normal ordered" Coulomb interaction

$$H' = \frac{1}{2} \int dx \int dx' \frac{e^2}{|x - x'|} N \left[\psi(x)^+ \psi(x')^+ \psi(x')\psi(x) \right] .$$

Besides the simplicity of the formulation, the normal ordering helps within the many body perturbation theory to simplify the Feynman diagram technique borrowed from the relativistic quantum field theory. We do not develop here this rather sophisticated technique, although it has broad and successful applications.

However, we have to stress that the similarity of the two theories is rather limited and from the strict mathematical point of view both stay on a shaky ground. Besides the questions about convergence, both are plagued by diverging integrals at high momenta ("ultraviolet divergences") as well as at very low momenta ("infrared divergences").

The ultraviolet problems are tackled in quite a different way. In the non-relativistic many body theories one restricts the integration domains by physically relevant momenta appropriate for the system one is modeling. In the case of the electrons in a crystal this is the Brillouin zone. Such kind of theories are called "cut-off" theories. Here the "bare" (unrenormalized i.e. in the absence of the interaction) electron and hole masses m_e, m_h, as well as the electron charge e are the experimental ones. On the contrary, in the quantum electrodynamics the divergences are eliminated by the so-called renormalization procedure admitting that actually the "bare" (input) parameters are infinite and the renormalized ones are fixed to give the finite physical electron charge and mass.

In what concerns the infrared divergences the problem has no simple solution due to the fact that in quantum electrodynamics one cannot differentiate the case without photon emission from the case of emission of several photons of zero momentum. Anyway, in the non-relativistic theory one has to take into account, that the Coulomb potential has no true Fourier transform, but only in the sense of the $\kappa \to 0$ limit of a Yukawa potential $\frac{1}{r}e^{-\kappa r}$.

Chapter 4
Phonons

The ions forming the attractively acting lattice are not rigid. At least their oscillations around the rigid lattice positions have to be taken into account. These oscillation may be quantized and interpreted as bosonic particles called acoustical or optical phonons with typical spectra. Even a classical description of the phonons gives rise to important predictions as the Lyddane-Sachs-Teller formula. In a quantum mechanical approach to optical transitions assisted by phonons one may understand the Franck-Condon effect.

4.1 Lattice Oscillations

Until now we considered the ions as fixed positive point-like charges, positioned at the sites $\mathbf{R} + \boldsymbol{\xi}_s$ ($s = 1 \ldots S$), where $\mathbf{R}$ is a vector of the Bravais lattice, while $\boldsymbol{\xi}_s$ shows the position of a given ion within the elementary cell containing S ions. Actually, the ions are free to move away from their equilibrium positions with deviations $\mathbf{u}(\mathbf{R}, s)$ (here still classical !). For small deviations one may develop the potential energy $\mathscr{U}$ of the lattice in a power series of these deviations. Up to a constant defining the ground state energy, the lowest term in this series must be quadratic, since the equilibrium corresponds to a stable minimum. Thus

$$\mathscr{U} = \frac{1}{2} \sum_{\mathbf{R},s} \sum_{\mathbf{R}',s'} \Phi_{ss'}^{\mu\nu}(\mathbf{R} - \mathbf{R}')u^{\mu}(\mathbf{R}, s)u^{\nu}(\mathbf{R}' s') + \ldots$$

where $\Phi_{ss'}^{\mu\nu}(\mathbf{R} - \mathbf{R}')$ is a real symmetric

$$\Phi_{ss'}^{\mu\nu}(\mathbf{R}) = \Phi_{s's}^{\nu\mu}(-\mathbf{R})$$

© Springer International Publishing AG, part of Springer Nature 2018
L. A. Bányai, *A Compendium of Solid State Theory*,
https://doi.org/10.1007/978-3-319-78613-1_4

positive defined matrix ($\Phi \geq 0$ means non-negative eigenvalues).

Since a constant shift does not change the energy

$$\sum_{\mathbf{R},s} \Phi_{ss'}^{\mu\nu}(\mathbf{R}) = \sum_{\mathbf{R},s} \Phi_{s's}^{\mu\nu}(\mathbf{R}) = 0 .$$

One gets the classical Lagrange function $L = \mathscr{T} - U$ of the lattice by considering also the kinetic energy of the lattice

$$\mathscr{T} = \frac{1}{2} \sum_{\mathbf{R},s} m_s \dot{\mathbf{u}}(\mathbf{R}, s)^2 ,$$

where m_s is the mass of the ion designed with the index s.

In terms of discrete Fourier transforms

$$\mathbf{u}(\mathbf{R}, s) = \frac{v}{(2\pi)^3} \int_{BZ} d\mathbf{q} e^{i\mathbf{q}\mathbf{R}} \tilde{\mathbf{u}}_s(\mathbf{q}); \qquad \tilde{\Phi}_{ss'}^{\mu\nu}(\mathbf{q}) = \frac{v}{(2\pi)^3} \int_{BZ} d\mathbf{q} e^{i\mathbf{q}\mathbf{R}} \Phi_{ss'}^{\mu\nu}(\mathbf{R})$$

retaining only the quadratic terms of the potential energy we have

$$L = \frac{v}{2(2\pi)^3} \int_{BZ} d\mathbf{q} \left\{ \sum_{s=1}^{S} \sum_{\mu} m_s \dot{\tilde{u}}_s^{\mu}(\mathbf{q})^* \dot{\tilde{u}}_s^{\mu}(\mathbf{q}) - \sum_{s,s'=1}^{S} \sum_{\mu,\nu} \tilde{u}_s^{\mu}(\mathbf{q})^* \tilde{\Phi}(\mathbf{q})_{ss'}^{\mu\nu} \tilde{u}_{s'}^{\nu}(\mathbf{q}) \right\}$$

with

$$\tilde{\Phi}(\mathbf{q})_{ss'}^{\mu\nu*} = \tilde{\Phi}(\mathbf{q})_{s's}^{\nu\mu} ; \qquad \tilde{\Phi}(\mathbf{q})_{ss'}^{\mu\nu*} = \tilde{\Phi}(-\mathbf{q})_{ss'}^{\mu\nu} ; \qquad \tilde{\Phi}(\mathbf{q}) \geq 0 .$$

Absorbing the mass factor by

$$\tilde{\eta}_s^{\mu}(\mathbf{q}) \equiv \sqrt{m_s} \tilde{u}_s^{\mu}(\mathbf{q}) ; \qquad \tilde{M}(\mathbf{q})_{ss'}^{\mu\nu} \equiv \frac{1}{\sqrt{m_s m_{s'}}} \tilde{\Phi}(\mathbf{q})_{ss'}^{\mu\nu}$$

we get

$$L = \frac{v}{2(2\pi)^3} \int_{BZ} d\mathbf{q} \left\{ \sum_{s=1}^{S} \sum_{\mu} \dot{\tilde{\eta}}_s^{\mu}(\mathbf{q})^* \dot{\tilde{\eta}}_s^{\mu}(\mathbf{q}) - \sum_{s,s'=1}^{S} \sum_{\mu,\nu} \tilde{\eta}_s^{\mu}(\mathbf{q})^* \tilde{M}(\mathbf{q})_{ss'}^{\mu\nu} \tilde{\eta}_{s'}^{\nu}(\mathbf{q}) \right\} ,$$

where the new matrix $\tilde{M}$ has the same properties of symmetry, realness and positiveness as $\tilde{\Phi}$.

This whole quadratic form may be brought to a diagonal one with the transformation, that diagonalizes the potential energy, with positive eigenvalues $\omega_\lambda(\mathbf{q})^2$ and one may see that the Lagrangian describes a sum of oscillators of unit mass with eigenfrequencies $\omega_\lambda(\mathbf{q}) = \omega_\lambda(-\mathbf{q})$.

The quantization of the lattice corresponds therefore to the quantization of these oscillators. As it is well known, this leads to a quantum mechanical Hamiltonian that may be formulated in terms of the creation and annihilation operators $b_{\lambda q}^{\dagger}$, $b_{\lambda q}$ and the respective oscillator frequencies $\omega_\lambda(\mathbf{q})$. It is convenient to impose cyclical boundary conditions in order to deal with a discrete spectrum having the proper number of degrees of freedom. Then the quantized lattice Hamiltonian is

$$H = \sum_{\mathbf{q} \in BZ} \sum_\lambda \hbar \omega_{\lambda \mathbf{q}} b_{\lambda \mathbf{q}}^{\dagger} b_{\lambda \mathbf{q}}$$

with the bosonic commutation relations

$$\left[b_{\lambda \mathbf{q}}, b_{\lambda' \mathbf{q}'} \right] = 0$$
$$\left[b_{\lambda \mathbf{q}}, b_{\lambda' \mathbf{q}'}^{\dagger} \right] = \delta_{\lambda, \lambda'} \delta_{\mathbf{q}, \mathbf{q}'}.$$

The quantized deviations from the equilibrium positions at their turn are

$$\mathbf{u}(\mathbf{R}, s) = \sum_{\mathbf{q}} \sum_\lambda e^{i \mathbf{q} \mathbf{R}} \sqrt{\frac{\hbar}{2 m_s \omega_{\lambda, \mathbf{q}}}} \chi_s^{(\lambda)}(\mathbf{q}) \left(b_{\lambda, \mathbf{q}} + b_{\lambda, -\mathbf{q}}^{\dagger} \right),$$

where $\chi_s^{(\lambda)}(\mathbf{q})$ are the orthonormalized eigenfunctions of the Matrix $\tilde{M}$

$$\sum_{\nu, s'} \tilde{M}(\mathbf{q})_{s, s'} \chi_{s'}^{\nu(\lambda)}(\mathbf{q}) = \omega_{\lambda, \mathbf{q}}^2 \chi_s^{\mu(\lambda)}(\mathbf{q}); \quad (\lambda = 1, \ldots 3S).$$

Due to the invariance against a common translation discussed before it follows, that

$$\sum_{s'=1}^{S} \tilde{M}(0)_{s, s'}^{\mu \nu} \sqrt{m_{s'}} = 0$$

and

$$\sum_{s=1}^{S} \sqrt{m_s} \tilde{M}(0)_{ss'}^{\mu \nu} = 0.$$

While the first equation shows that there are at least 3 eigenfrequencies that vanish at $\mathbf{q} = 0$, called acoustical modes, the second equation shows, that if an eigenvalue does not vanish in the origin, then the eigenstate must fulfill he relation

$$\sum_{s=1}^{S} \sqrt{m_s} \chi_s^{(\lambda)}(0) = 0; \quad (\lambda = 4, \ldots 3S - 3)$$

or

$$\sum_{s=1}^{S} m_s \mathbf{u}_s^{(\lambda)}(0) = 0; \qquad (\lambda = 4, \ldots 3S - 3).$$

This means that the center of mass by these oscillations, called optical modes, remains unchanged.

4.2 Classical Continuum Phonon-Model

In applications, it is useful to use a simplified continuum-model for phonons with a simple phonon spectrum. The local deviations $\mathbf{u}(\mathbf{r})$ are defined then in every space point $\mathbf{r}$.

The prototype classical Lagrange function for acoustical phonons is

$$L_{acc} = \frac{1}{2v} \int d\mathbf{r} m \sum_{\mu=1}^{3} \left\{ \dot{u}_\mu(\mathbf{r})^2 + c^2 \sum_{\nu=1}^{3} (\partial_\nu u_\mu(\mathbf{r}))^2 \right\}$$

or in Fourier transforms

$$L_{acc} = \frac{1}{2} \int d\mathbf{q} m \sum_{\mu=1}^{3} \left\{ \dot{\tilde{u}}_\mu(\mathbf{q})^* \dot{\tilde{u}}_\mu(\mathbf{q}) + c^2 q^2 \tilde{u}_\mu(\mathbf{q})^* \tilde{u}_\mu(\mathbf{q}) \right\} .$$

Obviously, here a linear acoustical phonon spectrum $\omega_{ac}(\mathbf{q}) = cq$ was assumed (with c being here the sound velocity in the medium).

To model optical phonons, one considers the classical Lagrange function

$$L_{opt} = \frac{1}{2v} \int d\mathbf{r} m \sum_{\mu=1}^{3} \left\{ \dot{\tilde{u}}_\mu(\mathbf{r})^2 - \omega_0^2 \tilde{u}_\mu(\mathbf{r}))^2 \right\}$$

or in Fourier transforms

$$L_{opt} = \frac{1}{2} \int d\mathbf{q} m \sum_{\mu=1}^{3} \left\{ \dot{\tilde{u}}_\mu(\mathbf{q})^* \dot{\tilde{u}}_\mu(\mathbf{q}) - \omega_0^2 \tilde{u}_\mu(\mathbf{q})^* \tilde{u}_\mu(\mathbf{q}) \right\} .$$

Here the optical phonon spectrum was taken just to be constant $\omega_{opt}(\mathbf{q}) = \omega_0$. In both cases one must take into account, that the total number of degrees of freedom (per volume!) should corresponds to that of the crystal and therefore one cuts of the wave vectors by the Debye wave vector ($q < q_{Debye}$) defined by

$$\frac{4\pi}{3} q_{Debye}^3 \equiv \frac{1}{v}$$

with v being the volume of the elementary cell.

4.2.1 Optical Phonons in Polar Semiconductors

The optical lattice deviations lead to local electric dipoles. This local polarization in the classical continuum model is given by

$$\mathbf{P}(\mathbf{r}) = \frac{\kappa e}{v} \mathbf{u}(\mathbf{r}) \ ,$$

where κ is the polarisability of the elementary cell. The corresponding polarization charge density is

$$\rho_{pol}(\mathbf{r}) = -\nabla \cdot \mathbf{P}(\mathbf{r}) = -\frac{\kappa e}{v} \nabla \cdot \mathbf{u}(\mathbf{r})$$

and the Poisson equation in the presence of a stationary external charge density $\rho_{ext}(\mathbf{r})$ looks as

$$\varepsilon_\infty \nabla^2 V(\mathbf{r}) = -4\pi \left(\rho_{pol}(\mathbf{r}) + \rho_{ext}(\mathbf{r}) \right)$$

with ε_∞ being the dielectric constant due to the electronic background.

Alternatively, for low frequencies it shall look as

$$\varepsilon_0 \nabla^2 V(\mathbf{r}) = -4\pi \rho_{ext}(\mathbf{r})$$

where ε_0 is the total dielectric constant. The notation stems from the assumption, that at high frequencies only the light electrons contribute to the dielectric properties, while for low frequencies both the electrons and ions contribute. As we shall see, relating the polarization charge density to the potential allows for the identification of the total dielectric constant.

In the inhomogeneous situation due to the presence of an external charge density the Lagrange function contains supplementary Coulomb terms

$$L = \frac{1}{2v} \int d\mathbf{r} m \sum_{\mu=1}^{3} \{ \dot{u}^\mu(\mathbf{r})^2 - \omega_0^2 u^\mu(\mathbf{r}))^2 \}$$
$$- \frac{1}{\varepsilon_\infty} \int d\mathbf{r} \int d\mathbf{r}' \frac{1}{|\mathbf{r} - \mathbf{r}'|} \left(\frac{1}{2} \nabla \mathbf{P}(\mathbf{r}) \nabla' \mathbf{P}(\mathbf{r}') - \nabla \mathbf{P}(\mathbf{r}) \rho_{ext}(\mathbf{r}') \right) \ .$$

Now let us split the local deviations $\mathbf{u}(\mathbf{r})$ into longitudinal and transverse modes

$$\mathbf{u}(\mathbf{r}) = \mathbf{u}_l(\mathbf{r}) + \mathbf{u}_t(\mathbf{r})$$

$$\mathbf{u}_l(\mathbf{r}) = -\nabla \int d\mathbf{r}' \frac{\nabla'\mathbf{u}(\mathbf{r}')}{4\pi |\mathbf{r} - \mathbf{r}'|}; \qquad \nabla \mathbf{u}_t = 0; \quad \nabla \mathbf{u}_l = \nabla \mathbf{u}.$$

One gets the equations of motion for these modes through the Euler-Lagrange equations by using the identity

$$\nabla^2 \frac{1}{|\mathbf{r}|} = -4\pi\,\delta(\mathbf{r})$$

and they look as

$$\frac{\partial^2}{\partial t^2}\mathbf{u}_t = -\omega_{TO}^2\mathbf{u}_t$$

$$\frac{\partial^2}{\partial t^2}\mathbf{u}_l = -\omega_{LO}^2\mathbf{u}_l + \frac{\kappa e}{m}\mathbf{E}_{ext}$$

with

$$\omega_{TO}^2 \equiv \omega_0^2; \qquad \omega_{LO}^2 = \omega_0^2 + \frac{4\pi\kappa^2 e^2}{\varepsilon_\infty vm}.$$

In a stationary regime (thermal equilibrium) in the presence of a stationary external charge density it follows

$$\mathbf{u}_l = \frac{\kappa e}{m\omega_{LO}^2}\mathbf{E}_{ext}$$

and

$$\rho_{pol} = -\frac{\kappa^2 e^2}{vm\omega_{LO}^2}\nabla\mathbf{E}_{ext} = -\frac{4\pi\kappa^2 e^2}{vm\omega_{LO}^2\varepsilon_\infty}\rho_{ext}; \qquad (\varepsilon_\infty \nabla\mathbf{E}_{ext} = 4\pi\rho_{ext}).$$

If one inserts this result into the Poisson equation one may identify

$$\frac{\varepsilon_\infty}{\varepsilon_0} = 1 - \frac{4\pi\kappa^2 e^2}{vm\omega_{LO}^2\varepsilon_\infty} = \frac{1}{\omega_{LO}^2}\left(\omega_{LO}^2 - \frac{4\pi\kappa^2 e^2}{\varepsilon_\infty vm}\right)$$

and one gets the Lyddane-Sachs-Teller relationship

$$\frac{\omega_{LO}}{\omega_{TO}} = \sqrt{\frac{\varepsilon_0}{\varepsilon_\infty}}.$$

Now let us introduce an arbitrary external time dependent electromagnetic field $\mathbf{E}_{ext}(t)$. Then the equation of motion looks as

$$\frac{\partial^2}{\partial t^2}\mathbf{u}(t) = -\omega_0^2\mathbf{u}(t) + \frac{\kappa e}{m}\mathbf{E}(t) \quad ,$$

whereby the electric field $\mathbf{E}(t)$ includes also the longitudinal field produced by the dipoles

$$\mathbf{E}(\mathbf{x}, t) = \mathbf{E}_{ext}(\mathbf{x}, t) + \nabla \int d\mathbf{x}' \frac{\nabla' P(\mathbf{x}', t)}{|\mathbf{x} - \mathbf{x}'|} \, .$$

Above we took into account that the interaction energy of our dipoles with the total electric field may be brought to the form $-\frac{\kappa e}{v} \int d\mathbf{x} \mathbf{u}(\mathbf{x}) \mathbf{E}(\mathbf{x}, t)$. After a Fourier transformation in the time variable t it follows for the frequency dependent susceptibility for both modes

$$\chi_{T,L}(k, \omega) = \frac{\alpha^2 e^2}{vm} \frac{1}{\omega_0^2 - \omega^2} \, ,$$

or for the frequency dependent dielectric function

$$\varepsilon(\omega)_{L,T} = \varepsilon_\infty \left(1 + \frac{4\pi \alpha^2 e^2}{\varepsilon_\infty vm} \frac{1}{\omega_0^2 - \omega^2} \right) = \varepsilon_\infty \frac{\omega_{LO}^2 - \omega^2}{\omega_{TO}^2 - \omega^2} \, .$$

4.2.2 Optical Eigenmodes

Starting from the above dielectric function we look now for possible electromagnetic eigenoscillations in such a medium. Since the system is homogeneous and isotropic, by using Fourier transforms it is useful to split the electromagnetic fields in their longitudinal, respectively transverse parts. The Maxwell equations for the magnetic $\mathbf{B}$, respectively electric field $\mathbf{E}$ and the polarization $\mathbf{P}$ look in their components as

$$B(k, \omega)_L = 0$$
$$B(k, \omega)_T = \frac{ck}{\omega} E(k, \omega)_T$$
$$\iota k \varepsilon_\infty E(k, \omega)_L = -4\pi \iota k P(k, \omega)_L + 4\pi \rho^{ext}(k, \omega)$$
$$(k^2 - \varepsilon_\infty \frac{\omega^2}{c^2}) E(k, \omega)_T = \frac{4\pi \omega^2}{c^2} P(k, \omega)_T - \frac{\iota 4\pi \omega}{c^2} j^{ext}(k, \omega)_T \, .$$

With

$$\mathbf{P}(\mathbf{k}, \omega) = \frac{1}{4\pi} \left(\varepsilon(\mathbf{k}, \omega) - \varepsilon_\infty \right) \mathbf{E}(\mathbf{k}, \omega)$$

and the previously deduced dielectric function it follows, that in the absence of external sources the homogeneous Maxwell equations still have non-vanishing solutions. For $\omega = \omega_{LO}$ there is a non-trivial longitudinal solution for any $k = |\mathbf{k}|$, while non-trivial transverse solutions exist only for ω, k pairs that are solutions of the equation

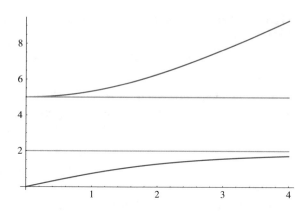

Fig. 4.1 Optical eigenmodes

$$k^2 - \varepsilon_\infty \frac{\omega^2}{c^2} \frac{\omega_{LO}^2 - \omega^2}{\omega_{TO}^2 - \omega^2} = 0$$

shown in Fig. 4.1. Such transverse fields are propagating mixed photon and LO-phonon modes.

4.2.3 The Electron-Phonon Interaction

4.2.3.1 The Franck-Condon Effect

Let us suppose, that one has two localized electronic states of energies ε_1 and ε_2 on the same site and one of them is electrically neutral while the other one is electrically charged. In a polar semiconductor, this local charge density is coupled to the optical phonons. The energy of the system of electron and phonons (seen as a classical oscillator with coordinate Q) may be characterized schematically by the potential energies of the two states

$$E_1(Q) = \varepsilon_1 + \frac{1}{2}m(\dot{Q}^2 + \omega_{LO}^2 Q^2)$$
$$E_2(Q) = \varepsilon_2 + \frac{1}{2}m(\dot{Q}^2 + \omega_{LO}^2 Q^2) - gQ \ .$$

The energy of the second state may be rewritten as

$$E_2(Q) = \varepsilon_2 + \frac{1}{2}m\left(\dot{Q}^2 + \omega_{LO}^2(Q - Q_0)^2 + \omega_{LO}^2 Q_0^2\right),$$

where $m\omega_{LO}^2 Q_0 = g$. This means that the potential energy of the phonons gets shifted.

Fig. 4.2 Stokes shift

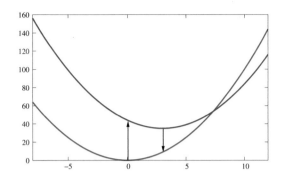

Now by photon absorption the system may undergo a transition from the minimum of the lower lying state on the higher branch outside the minimum of that branch. (The photon momentum may be neglected by these optical transitions!) Thereafter, follows a thermal relaxation (by emission of acoustical phonons) into the minimum of the upper branch and only later may follow a slower photon emission onto the lower branch again outside the minimum. This is the essence of the Franck-Condon effect showing a difference between the absorption and emission spectra (Stokes shift). It is illustrated in Fig. 4.2.

4.2.3.2 The Quantized Interaction of Electrons with Phonons

The interaction Hamiltonian of the electrons with the phonons is considered to be linear in the quantized lattice deviations, conserving the number of electrons and in the continuum model conserving also the momenta

$$H_{int} = \frac{1}{\sqrt{\Omega}} \sum_{\mathbf{q}} g_q a_{\mathbf{k}}^+ a_{\mathbf{k-q}} (b_{\mathbf{q}} + b_{-\mathbf{q}}^+).$$

Here Ω is the volume and the discretized wave vector $\mathbf{q}$ is assumed to be smaller than the Debye wave vector $q < q_D$.

In the case of the optical phonons, analogously to the discussion in the frame of the classical continuum model one starts from the Coulomb interaction between the electron (or hole) charge density

$$\rho(\mathbf{x}) \equiv \pm e \psi^+(\mathbf{x}) \psi(\mathbf{x})$$

and the polarization charge density due to the optical deformation

$$-\frac{1}{\varepsilon_\infty} \int d\mathbf{x} \int d\mathbf{x}' \frac{\nabla \mathbf{P}(\mathbf{x}) \rho(\mathbf{x}')}{|\mathbf{x} - \mathbf{x}'|} .$$

Inserting the expression of the polarization in terms of the quantized lattice deviations one gets

$$g_q^2 = \alpha \frac{4\pi \hbar \, (\hbar \omega_0)^{3/2}}{(2m_e)^{1/2} \, q^2} \, ,$$

where instead the coefficient $\frac{4\pi e^2 \kappa^2}{v m \omega_{LO}^2 \varepsilon_\infty}$ one has introduced the dimensionless constant

$$\alpha = \frac{e^2}{\hbar} \left(\frac{m_e}{2\hbar \omega_0} \right)^{1/2} \left(\frac{1}{\varepsilon_\infty} - \frac{1}{\varepsilon_0} \right) .$$

The presence of the electron mass here is spurious. The coupling constant g_q itself does not depend on the electron mass.

In the case of the acoustic phonons one considers mostly the deformation potential model. This starts from the assumption, that a slowly varying deformation in an isotropic medium causes a local variation of the band gap

$$\delta E_g(\mathbf{x}) \sim \nabla \mathbf{u}(\mathbf{x}) .$$

This leads to

$$g_q = G\sqrt{\hbar \omega_q} \, ,$$

where $\omega_q = c|\mathbf{q}|$ and G is a constant specific for the considered crystal. Of course, both models are highly idealized, but nevertheless, quite successful in predicting experimental phenomena.

Chapter 5
Transport Theory

The understanding of transport phenomena transcends the quantum-mechanical treatment of isolated systems. Irreversibility and dissipation due to the interaction with another macroscopic system (bath) play here a central role. For their understanding a simple classical, but exactly solvable polaron model is helpful. On the other hand, for all practical purposes an intuitive, however mathematically less sound treatment serves by introducing irreversibility "by hand" in the Boltzmann, Master or rate equations. In this frame the conductivity of electrons scattered on a thermal bath is discussed. Two basic arts of conduction are discussed. That of scattering of freely moving band electrons, typical for metals and semiconductors, as well as that of localized electrons hopping from site to site, typical for disordered systems, like impurity states or amorphous semiconductors. The transverse magneto-resistance in ultra-high magnetic fields also may be understood within this last model.

5.1 Non-equilibrium Phenomena

Most experiments on solid materials concern their electrical, optical, thermal and mechanical properties. Here we shall discuss only the first two topics, that imply interaction with external electromagnetic fields. Under such an external influence the solid system is brought into a non-equilibrium state. This may be a stationary state or a transitory time dependent evolution towards equilibrium. Classical or quantum mechanics of finite closed systems being time-reversal invariant cannot properly describe such irreversible processes. The origin of the irreversibility of real physical

© Springer International Publishing AG, part of Springer Nature 2018
L. A. Bányai, *A Compendium of Solid State Theory*,
https://doi.org/10.1007/978-3-319-78613-1_5

systems lies in the fact, that they are actually open systems connected with some thermostat to which they transfer (dissipate) energy. In other words, they are parts of a bigger system. The reversibility even of isolated systems may be however pushed far away in time due to the macroscopic size of the system. Therefore, the thermodynamic limit plays an essential role in the treatment of non-equilibrium phenomena.

The traditional theoretical treatment of non-equilibrium phenomena starts however directly by assuming some simple, intuitive assumptions leaving their proof apart. In the classical non-quantum-mechanical treatment such a useful tool is the Boltzmann equation for the statistical evolution of classical particles having a given velocity and position. We have already discussed, that electrons in a periodical lattice maybe also assimilated with free electrons having an effective mass. In this sense, one may use this model. On the other hand, one may encounter situations, where the quantum-mechanical properties are essential. In that case Master and rate equations are used to describe the evolution due to transitions between quantum mechanical states.

5.2   Classical Polaron in Electric Field

For a better understanding of the origin of irreversibility we describe here a classical solvable model of an electron interacting with classical continuum LO-phonons (polaron). The Froehlich coupling of a classical electron to a continuum of LO-phonons

$$-\frac{e}{\varepsilon_\infty} \int d\mathbf{x} \frac{\nabla \mathbf{P}(\mathbf{x})}{|\mathbf{x} - \mathbf{r}(t)|}$$

is just the ordinary Coulomb interaction between the electron of coordinate $\mathbf{r}(t)$ and the polarization charge caused by the lattice deviation field $\mathbf{u}(\mathbf{x})$. This polarization is proportional to $\mathbf{u}(\mathbf{x})$ (see Chap. 4)

$$\mathbf{P}(\mathbf{x}) = \frac{\kappa e}{v} \mathbf{u}(\mathbf{x}) .$$

However, before further proceeding, one is compelled to regularize the continuum phonon model in such a way as to accommodate its degrees of freedom with those of a discrete lattice, as we discussed earlier. Otherwise the model is mathematically not well-defined (contains diverging entities). This may be achieved also by choosing instead the Coulomb potential $\frac{e}{r}$ another one $V(r)$, whose Fourier transform $\tilde{V}(\mathbf{q})$ vanishes above the Debye wave vector q_D.

The coupled Euler-Lagrange equations for the electron coordinate $\mathbf{r}(t)$ in the presence of a longitudinal, homogeneous, constant electric field $\mathbf{E}$ (for sake of simplicity acting only on the electron) and the longitudinal phonon mode $d(\mathbf{x}, t) \equiv \nabla \mathbf{u}(\mathbf{x}, t)$ are

$$m_e \ddot{\mathbf{r}}(t) = e\mathbf{E} - \frac{\kappa e^2}{\varepsilon_\infty} \int d\mathbf{x} d(\mathbf{x}, t) \nabla V(\mathbf{x} - \mathbf{r}(t))$$

$$\ddot{d}(\mathbf{x}, t) + \omega_{LO}^2 d(\mathbf{x}, t) = -\frac{\kappa}{m\varepsilon_\infty} \nabla^2 V(\mathbf{x} - \mathbf{r}(t)) \,.$$

These equations are describing a classical mechanical system and are perfectly invariant against time reversal. We shall study the motion of this entire system, but looking in detail only on what happens to the electron.

Let us consider that at some initial time instant $t = 0$ the phonons and the electron were in rest

$$d(\mathbf{x}, 0) = 0; \qquad \dot{d}(\mathbf{x}, 0) = 0; \qquad \dot{\mathbf{r}} = 0$$

and under the influence of the applied electric field they both start to move.

The phonon equation may be solved formally

$$d(\mathbf{x}, t) = -\frac{\kappa e^2}{m\varepsilon_\infty} \int_0^t dt' \sin \omega_{LO}(t - t') \nabla^2 V(\mathbf{x} - \mathbf{r}(t'))$$

and inserting it in the equation motion of the electron we get a non-linear, non-local in time and coordinate, but closed equation for the electron coordinate

$$\ddot{\mathbf{r}}(t) = \frac{e}{m_e} \mathscr{E} + C \int_0^t dt' \sin \omega_{LO}(t - t') \frac{\partial}{\partial \mathbf{r}(t)} \mathscr{V}(\mathbf{r}(t) - \mathbf{r}(t')) \,,$$

where the kernel $\mathscr{V}(\mathbf{r})$ is defined by

$$\mathscr{V}(\mathbf{r}) \equiv -\frac{1}{4\pi} \int d\mathbf{x} V(\mathbf{r} - \mathbf{x}) \nabla^2 V(\mathbf{x})$$

and the constant is

$$C \equiv \frac{4\pi \kappa^2 e^4}{mm_e \varepsilon_\infty^2} \,.$$

It is convenient to consider a simple Debye regularization of the potential by the choice

$$V(r) = \frac{1 - e^{-q_D r}}{r}$$

having the Fourier transform $\tilde{V}(q) = \frac{4\pi q_D^2}{q^2(q^2 + q_D^2)}$. This implies a smooth vanishing at $q \gg q_D$, but other possible choices do not modify qualitatively the results. With this choice the kernel looks explicitly as

$$\mathscr{V}(r) = q_D \left(\frac{1 - e^{-q_D r}}{q_D r} - \frac{1}{2} e^{-q_D r} \right).$$

In order to further simplify the problem, we consider just a co-linear motion of the electron $\dot{\mathbf{r}}(0) \parallel \mathbf{E}$. Then we have a scalar equation

$$\frac{d^2 x(t)}{dt^2} = \mathscr{E} + C \int_0^t dt' \sin(t - t') \frac{\partial}{\partial x(t)} \mathscr{V}(|x(t) - x(t')|)$$

to solve. Here we have chosen q_D^{-1} as length unit and ω_{LO}^{-1} as time unit, while $\mathscr{E} \equiv \frac{e q_D}{m_e} E$.

As simple as this equation may look, one must use a computer to solve it numerically. Nevertheless, we may see analytically the asymptotic behavior of the solution.

Let us find out whether an expected asymptotic uniform motion in the field (according to Ohm's law for weak fields)

$$x(t) \approx vt \qquad (t \to \infty)$$

may result. Obviously, for this sake one has to fulfill the condition

$$0 = \mathscr{E} - C \int_0^\infty dt \sin t \frac{\partial}{\partial vt} \mathscr{V}(vt)$$

or explicitly the transcendent equation

$$\frac{\mathscr{E}}{C} = \mathscr{K}(v)$$

with

$$\mathscr{K}(v) \equiv \frac{1}{2v^2} \left\{ \ln(1 + v^2) - \frac{1}{1 + v^{-2}} \right\}.$$

Since the function $\mathscr{K}(v)$ is positive, but bounded from above

$$0 \le \mathscr{K}(v) \le 1.08108,$$

steady asymptotical motion may occur only below the critical field strength

$$\mathscr{E}_c = 1.08108 \times C.$$

A numerical solution of the equation of motion shows in Fig. 5.1 the velocity $v(t)$ for two field strengths, one below and another above the critical one. In the first case, a steady motion sets in (Ohm's law), while in the second case the motion gets accelerated (break-through). The accelerated motion in the presence of the electric field agrees with the requirement of time-reversal for the electron-field system (ignoring the phonons), but the steady one is not. It may result only due to "non-mechanical" friction (damping).

Fig. 5.1 The velocity $v(t)$ for $E < E_c$ (red) and $E_c > E_c$ (green)

Unfortunately, besides this simple classical model, even by numerical methods, no solutions may be found for realistic classical or quantum mechanical evolution of electrons coupled to another macroscopic system ("thermal bath"). Therefore, one uses instead certain assumptions about irreversible terms in the equations describing the evolution. While no consistent mathematical proof of those assumptions is available, they proved themselves by very good qualitative and quantitative predictions. In what follows we shall describe these approaches.

5.3 The Boltzmann Equation

In the frame of the classical statistical mechanics a system of point-like particles (inside a Volume Ω) is characterized by the probability density $p(\mathbf{v}, \mathbf{x}, t)$ to find at time instant t a particle with velocity $\mathbf{v}$ at the space-point $\mathbf{x}$, normalized as

$$\int d\mathbf{v} \int_{\Omega} d\mathbf{x}\, p(\mathbf{v}, \mathbf{x}, t) = 1 .$$

The classical Liouville equation for the motion in the presence of electric $\mathbf{E}(\mathbf{x}, t)$ and magnetic fields $\mathbf{B}(\mathbf{x}, t)$ is

$$\left(\frac{\partial}{\partial t} + \mathbf{v}\frac{\partial}{\partial \mathbf{x}} + \frac{1}{m}\mathbf{F}(\mathbf{x}, t)\frac{\partial}{\partial \mathbf{v}} \right) p(\mathbf{k}, \mathbf{x}, t) = 0 ,$$

where $\mathbf{F} = e(\mathbf{E} + \frac{1}{c}\mathbf{v} \times \mathbf{B})$ is the Lorenz force. The meaning of this equation is that along the trajectory the probability density is conserved. Since the charged particles themselves are sources of fields, one must include their contribution. If the velocity of the particles is very low one may ignore their contribution to the magnetic field and for sake of simplicity we ignore for instant any magnetic field. For the electric field, in a self-consistent manner, one considers the Poisson equation

$$\varepsilon_0 \nabla \mathbf{E}(\mathbf{x}, t) = 4\pi \left(\rho(\mathbf{x}, t) + \rho_{ext}(\mathbf{x}, t) \right) ,$$

where the particle charge density is

$$\rho(\mathbf{x}, t) \equiv eN \left(\int d\mathbf{v} p(\mathbf{v}, \mathbf{x}, t) - \frac{1}{\Omega} \right)$$

and ρ_{ext} is the external charge density. Here we have included also the contribution of a compensating uniform positive charge in the volume Ω, that keeps the system stable.

Up to this point we are still in the frame of a reversible mechanical description, however, we may introduce a term describing collisions either on some disordered potential, on phonons in equilibrium, or between the particles themselves, which gives rise to irreversibility.

$$\left(\frac{\partial}{\partial t} + \mathbf{v} \frac{\partial}{\partial \mathbf{x}} + \frac{1}{m} \mathbf{F}(\mathbf{x}, t) \frac{\partial}{\partial \mathbf{v}} \right) p(\mathbf{v}, \mathbf{x}, t) = \frac{\partial p(\mathbf{v}, \mathbf{x}, t)}{\partial t} \Big|_{coll}$$

In solids, at relatively low particle densities (semiconductors) one considers mostly the scattering on external disturbances (disorder, phonons) and the collision term is obtained from statistical loss-gain considerations

$$\frac{\partial p(\mathbf{v}, \mathbf{x}, t)}{\partial t} \Big|_{coll} = - \int d\mathbf{v}' \left[w(\mathbf{v}, \mathbf{v}') p(\mathbf{v}, \mathbf{x}, t) - w(\mathbf{v}', \mathbf{v}) p(\mathbf{v}', \mathbf{x}, t) \right]$$

with transition rates $w(\mathbf{v}, \mathbf{v}')$ characterizing the abrupt velocity changes due to collisions. In solid-state theory, these transition rates are borrowed from the "golden rule" of quantum-mechanics and therefore, we can speak about it as a quasi-classical description. The transition rates are supposed to satisfy the *detailed balance relation* (its justification will be discussed later)

$$w(\mathbf{v}, \mathbf{v}') = e^{\beta(e(\mathbf{v}) - e(\mathbf{v}'))} w(\mathbf{v}', \mathbf{v})$$

implying the kinetic energy $e(\mathbf{v}) = \frac{mv^2}{2}$ before and after the collision. In the peculiar case of elastic scattering, of course, the energy is conserved, while by scattering on phonons it changes due to phonon absorption or emission.

The phonon system at the inverse temperature β acts as a thermal reservoir and (if the external sources are not time dependent and with vanishing boundary conditions!) obviously the stable stationary solution is the Maxwell equilibrium distribution

$$p_0(\mathbf{v}, \mathbf{x}, t) = \frac{1}{\Omega Z} e^{-\beta(\frac{mv^2}{2} + U(\mathbf{x}))} ,$$

with $U(\mathbf{x})$ being the s.c. equilibrium potential.

These equations conserve the total number of particles i.e. the normalization

$$\frac{\partial}{\partial t} \int d\mathbf{v} \int_\Omega d\mathbf{x} d\mathbf{x} p(\mathbf{v}, \mathbf{x}, t) = 0 .$$

In the elastic case, also the total average energy is conserved.

$$\frac{\partial}{\partial t} \left[N \int d\mathbf{v} d\mathbf{x} p(\mathbf{v}, \mathbf{x}, t) \, (e(v) + U_{ext}(\mathbf{x})) + \frac{1}{2} \int d\mathbf{x} \int d\mathbf{x}' \frac{\rho(\mathbf{x}, t)\rho(\mathbf{x}', t)}{\varepsilon_0 |\mathbf{x} - \mathbf{x}'|} \right] = 0 .$$

However, one may conceive also flow situations by considering appropriate boundary conditions.

5.3.1 Classical Conductivity

Let us now consider a simple choice of the transition rates ("instant relaxation") by

$$w(\mathbf{v}', \mathbf{v}) = \frac{1}{\tau} f_0(e(v))$$

where τ is a constant "relaxation time" and

$$f_0(e) \equiv \frac{1}{Z} e^{-\beta e}; \quad e(v) = \frac{mv^2}{2}; \quad Z \equiv \int d\mathbf{v} e^{-\beta \frac{mv^2}{2}} ,$$

which obviously satisfies the balance relation. We shall look now for a spacial homogeneous flow situation in an infinite medium with a constant, homogeneous magnetic field $\mathbf{B}$ and an alternating homogeneous electric field $\mathbf{E} e^{i\omega t}$. (Since the uniform charge density of the particles is supposed to be compensated by a background positive charge, in this configuration the external field coincides with the real field in the system!)

In this case for the velocity distribution

$$f(\mathbf{v}, t) = \int d\mathbf{x} p(\mathbf{v}, \mathbf{x}, t)$$

the Boltzmann equation takes a simplified form

$$\left(\frac{\partial}{\partial t} + \frac{e}{m} \left(\mathbf{E} e^{i\omega t} + \frac{1}{c} \mathbf{v} \times \mathbf{B} \right) \frac{\partial}{\partial \mathbf{v}} \right) f(\mathbf{v}, t) = -\frac{f(\mathbf{v}, t) - f_0(e(v))}{\tau} .$$

The right hand side is just an ideal relaxation term with the relaxation time τ. In the stationary flow state, the deviation of the solution from equilibrium one

$$f(\mathbf{v},t) = f_0(e(v)) + g_1(\mathbf{v})e^{\iota\omega t} + \dots$$

should be at least of first order in the electric field. Retaining only terms of first order in the field we get

$$\left(\iota\omega + \frac{1}{\tau} + \frac{e}{mc}(\mathbf{v}\times\mathbf{B})\frac{\partial}{\partial\mathbf{v}}\right)g_1(\mathbf{v}) = -e\mathbf{E}\mathbf{v}\frac{\partial f_0(e(v))}{\partial e(v)} \ .$$

To solve this equation, we have to inverse the operator

$$\hat{A} \equiv \iota\omega + \frac{1}{\tau} + \frac{e}{mc}(\mathbf{v}\times\mathbf{B})\frac{\partial}{\partial\mathbf{v}} \ .$$

This may be easily done by knowing its eigenfunctions and eigenvalues. Let us choose the magnetic field along the z axis $\mathbf{B} = (B,0,0)$. Then we have the following eigenstates:

$$\begin{array}{lll}(\mathbf{e_x v}) & with\ eigenvalue & \iota\omega + \frac{1}{\tau} + \omega_c v_y\\ (\mathbf{e_y v}) & with\ eigenvalue & \iota\omega + \frac{1}{\tau} - \omega_c v_x\\ (\mathbf{e_z v}) & with\ eigenvalue & \iota\omega + \frac{1}{\tau}\end{array}$$

where $\omega_c = \frac{eB}{mc}$ is the cyclotron frequency. Now one may compute the average electric current density

$$\langle\mathbf{j}\rangle \equiv e\langle n\rangle \int d\mathbf{v}\,\mathbf{v} f(\mathbf{v})$$

(here $\langle n\rangle = \frac{\langle N\rangle}{\Omega}$ is the average particle density) to get the conductivity tensor $\sigma_{\mu v}(\omega)$ (defined by $\langle j_\mu\rangle = \sigma_{\mu v}E_v$):

$$\sigma_{zz}(\omega,\omega_c) = \frac{\langle n\rangle e^2}{m}\frac{1}{\iota\omega + \frac{1}{\tau}}$$

$$\sigma_{xx}(\omega,\omega_c) = \sigma_{xx}(\omega,\omega_c) = \frac{\langle n\rangle e^2}{2m}\left(\frac{1}{\iota(\omega-\omega_c)+\frac{1}{\tau}} + \frac{1}{\iota(\omega+\omega_c)+\frac{1}{\tau}}\right)$$

$$\sigma_{xy}(\omega,\omega_c) = -\sigma_{yx}(\omega,\omega_c) = \frac{\langle n\rangle e^2}{2m}\left(\frac{1}{\iota(\omega-\omega_c)+\frac{1}{\tau}} - \frac{1}{\iota(\omega+\omega_c)+\frac{1}{\tau}}\right)$$

$$\sigma_{xz}(\omega,\omega_c) = \sigma_{zx}(\omega,\omega_c) = \sigma_{yz}(\omega,\omega_c) = \sigma_{zy}(\omega,\omega_c) = 0$$

Of course, this is an oversimplified model, but nevertheless it shows the role of the relaxation time τ, the phenomenon of cyclotron resonance at $\omega \Rightarrow \omega_c$, as well as the vanishing of some components of the conductivity tensor.

5.4 Master and Rate Equations

5.4.1 Master Equations

Obviously, a quasi-classical description as in the previous subsection is of very limited validity. From a quantum-mechanical point of view, simultaneous coordinate and momentum are not the proper characterization of a particle. Actually, some quantum numbers characterize the state. Therefore within the same statistical gain-loss philosophy one uses the so called Master and rate equations.

The statistical state ("mixed state") of a quantum mechanical system may be defined by the probability $P_n(t)$ to find the system at instant t in the eigenstate $|n>$ of an "unperturbed" Hamilton operator H_0. The time evolution of this statistical system is due to some interaction H' causing transitions between the unperturbed states $|n>$ and one assumes, that the transition rates $W_{nn'}$ are given by the "golden rule" of adiabatic perturbation theory:

$$W_{nn'} \equiv \lim_{\Delta t \to \infty} \frac{1}{\Delta t} \left| < n' |U(\Delta t)| n > \right|^2$$

with $U(\Delta t)$ being the unitary evolution operator during the time interval Δt.

The Master equation looks then as

$$\frac{\partial}{\partial t} P_n(t) = - \sum_{n'} (P_n(t) W_{nn'} - P_{n'}(t) W_{n'n}) \equiv - \sum_{n'} A_{nn'} P_{n'}(t) .$$

The general properties of the transitions rates emerge from the "golden rule" definition. The transition rates are obviously non-negative and satisfy the generalized balance property emerging from the unitarity of U:

$$W_{nn'} \geq 0; \qquad \sum_{n'} (W_{nn'} - W_{n'n}) = 0 .$$

In virtue of these properties for the solution of the Master equation it follows that:

(i) normalization is conserved $\frac{\partial}{\partial t} \sum_{n'} P_{n'}(t) = 0$,

(ii) positivity is conserved i.e. $P_n(t) \geq 0$ for any $t > 0$ if $P_n(0) \geq 0$. (it emerges from the fact that $P_n(t) = 0$ implies $\frac{\partial}{\partial t} P_n(t) \geq 0$).

(iii) the energy is conserved due to the admitted adiabatic coupling-decoupling of the interaction,

(iv) the evolution leads to a stable equilibrium state (irreversibility).

The formal solution is

$$P(t) = e^{-At} P(0)$$

and the operator A has non-negative eigenvalues.

Indeed, if $\xi_n = x_n + \imath y_n$ is a complex number, then

$$\Re\left\{\sum_{n,n'}\xi_n^* A_{nn'}\xi_{n'}\right\} = \Re\left\{\sum_{n,n'} W_{nn'}\xi_n^*\,(\xi_n - \xi_{n'})\right\}$$

$$= \frac{1}{2}\sum_{n,n'} W_{n'n}\left[(x_n - x_{n'})^2 + (y_n - y_{n'})^2\right] \geq 0\ ,$$

where we used also the balance property. If a general connectivity occurs, i.e. any state may be achieved from any state through successive transitions, then there is a single eigenstate withe eigenvalue 0 and its components are all equal ($\xi_n^0 = const.$). It follows therefore

$$\lim_{t\to\infty} P(t) = P^0$$

where $P_n^0 = const.$ and

$$P(t) = e^{-At}(P(0) - P^0) + P^0\ .$$

Such an evolution is called a Markowian process.

Let us now assume, that we have a system in contact with a macroscopic thermostat, whose state may be admitted remaining unchanged in thermal equilibrium Then the probability of the total state (system plus thermostat) may be admitted to be

$$P_{n\alpha}(t) = \mathscr{P}_n(t)\mathscr{P}_\alpha^0;\qquad \mathscr{P}_\alpha^0 = \frac{e^{-\beta E_\alpha}}{\sum_{\alpha'} e^{-\beta E_{\alpha'}}}\ ,$$

where the quantum numbers α characterize the states of the thermostat.
From the Master equation for $P_{n\alpha}(t)$ it follows

$$\frac{\partial}{\partial t}\mathscr{P}(t) = -\mathscr{A}\,\mathscr{P}(t);\qquad \mathscr{A}_{nn'} = \delta_{nn'}\sum_{n''} \mathscr{W}_{nn'} - \mathscr{W}_{n'n}$$

From the energy conservation ($E_n + E_\alpha = E_{n'} + E_{\alpha'}$) and the balance property of the total system follows the balance property of the averaged transition rates $\mathscr{W}_{nn'}$

$$\mathscr{W}_{nn'} \equiv \sum_{\alpha\alpha'} \mathscr{P}_\alpha^0 W_{n\alpha,n'\alpha'}$$

$$\sum_{n'}\left(\mathscr{W}_{nn'} - e^{\beta(E_n - E_{n'})}\mathscr{W}_{n'n}\right) = 0\ .$$

The operator $\mathscr{A}$ again hat (under a similar connectivity condition) only eigenvalues with non-negative real part, whereas the only eigenstate corresponding to the zero eigenvalue is

$$P_n^0 = \frac{e^{-\beta E_n}}{\sum_{n'} e^{-\beta E_{n'}}}\ .$$

5.4.2 Rate Equations

Let us consider the case, where the index $n \equiv \{n_1, n_2, \ldots\}$ corresponds to the set of occupation numbers of some one-electron states $|i>$, $(i = 1, 2, \ldots)$. The energy (without any interaction between the electrons!) is

$$E_n = \sum_i e_i n_i$$

and let us consider that through interaction with some thermal bath (for example, phonons) transitions occur. In lowest order of adiabatic perturbation theory the transition rate $\mathscr{W}_{nn'}$ is the sum of one-electron transition rates

$$\mathscr{W}_{nn'} = \sum_{i,j} w_{ij} n_i (1 - n'_j) \Pi_{l \neq i, j} \delta_{n_l, n'_l} .$$

Within this lowest order perturbation approximation even a "detailed balance relation"

$$w_{ij} = w_{ji} e^{\beta(e_i - e_j)}$$

occurs. Then the evolution of the average occupation number $\langle n_i(t) \rangle$ of a state i

$$\langle n_i(t) \rangle \equiv \sum_n n_i P_n(t)$$

looks as

$$\frac{\partial}{\partial t} \langle n_i(t) \rangle = -\sum_{n,n'} \mathscr{W}_{nn'} (n_i - n'_i) P_n(t) = -\sum_{i,j} \left[w_{ij} \langle n_i(1 - n_j) \rangle - w_{ji} \langle n_j(1 - n_i) \rangle \right]$$

As we see the equation is not closed, it includes the next average $\langle n_i n_j \rangle$. However, if one admits the approximation

$$\langle n_i n_j \rangle \approx \langle n_i \rangle \langle n_j \rangle$$

one gets a closed non-linear rate equation for the fermions

$$\frac{\partial}{\partial t} \langle n_i(t) \rangle = -\sum_{i,j} \left[w_{ij} \langle n_i(t) \rangle \left(1 - \langle n_j(t) \rangle \right) - w_{ji} \langle n_j(t) \rangle \left(1 - \langle n_i(t) \rangle \right) \right] .$$

Here obviously the Pauli principle is respected only on average. The average occupation numbers remain in the interval $[0, 1]$ if they were in this interval and under the previously discussed connectivity condition the stable asymptotic solution is the Fermi function

$$f(\varepsilon_i) = \frac{1}{e^{\beta(\varepsilon_i - \mu)} + 1} \, .$$

The new constant μ here is the chemical potential fixed by the average total number of the fermions, which is conserved by the transitions.

5.4.3 Hopping Transport

A peculiar transport mechanism may occur in solids, when conduction occurs through jumping between localized states. This is a typical situation in disordered semiconductors or, as we shall see later, even in crystalline semiconductors under ultra-strong magnetic fields.

Let us consider here a system of neutral fermions on localized states on an arbitrary disordered or ordered lattice points $\mathbf{r}_i$. The state of the fermion system is therefore described by the set of occupation numbers $n_i = 0, 1$ of the localized states having the one-particle energies ε_i. The total energy is

$$E_n = \sum_i \varepsilon_i n_i \, .$$

We shall describe the time-evolution of this system again by a rate equation

The transition rates w_{ij} are assumed to obey the balance property of the preceding Subsection. The only global observable we may consider beside the total energy, is the average particle density

$$\langle n(\mathbf{x}, t) \rangle = \sum_i \langle n_i(t) \rangle \delta(\mathbf{x} - \mathbf{x}_i) \, .$$

In the case of charged particles (electrons) the particle density would define also the electric potential created by them.

Let us assume, that we want to describe the evolution of a state which is not far away from equilibrium described by the Fermi function $f_i \equiv f(\varepsilon_i)$ i.e.

$$\langle n_i(t) \rangle = f_i + \delta \langle n_i(t) \rangle$$

with $\delta \langle n_i(t) \rangle$ very small. Then we may linearize the rate equation as

$$\frac{\partial}{\partial t} \delta \langle n(t) \rangle = -\Gamma \delta \langle n(t) \rangle$$

with

$$\Gamma_{ij} \equiv \left(\delta_{ij} \sum_l \mathcal{W}_{il} - \mathcal{W}_{ij} \right) \frac{1}{f_i(1 - f_i)}$$

and

$$\mathcal{W}_{ij} = f_i(1 - f_j)w_{ij} = \mathcal{W}_{ji} \ .$$

The real matrix Γ is hermitian in the scalar product

$$(\psi, \phi) \equiv \sum_i \psi_i^* \phi_i \frac{1}{f_i(1 - f_i)}$$

and non-negative since

$$(\psi, \Gamma\psi) = \frac{1}{2} \sum_{i,j} \left| \frac{\psi_i}{f_i(1 - f_i)} - \frac{\psi_j}{f_j(1 - f_j)} \right|^2 \mathcal{W}_{ij} \ge 0 \ .$$

Assuming connectivity of the lattice, there is just one null-eigenvector

$$\xi_i = \frac{f_i(1 - f_i)}{\sum_j \sqrt{f_j(1 - f_j)}} \ .$$

Therefore, the deviation from the equilibrium solution

$$\delta\langle n(t)\rangle = e^{-\Gamma t}\delta\langle n(0)\rangle$$

always vanishes asymptotically, $\delta\langle n(t)\rangle \to 0$ as $t \to \infty$. The null-eigenvector plays no role here, since the total particle number conservation implies

$$\sum_i \delta\langle n(t)\rangle = 0$$

i.e. $\delta\langle n(t)\rangle$ has no components in the direction of ξ

$$(\xi, \delta\langle n(t)\rangle) = 0 \ .$$

5.4.4 Hopping Diffusion on a Periodic Cubic Lattice

One can find a simple explicit solution for the hopping on a periodical cubic Bravais lattice of states of equal energies with translation invariant transition rates $w(\mathbf{R}, \mathbf{R}') \equiv w(\mathbf{R} - \mathbf{R}') > 0$ of finite range. The detailed balance here looks as

$$w(\mathbf{R} - \mathbf{R}') = w(\mathbf{R}' - \mathbf{R}).$$

The discrete Fourier transformation $\delta\tilde{n}(\mathbf{k}) \equiv \sum_{\mathbf{R}} \delta\langle n_{\mathbf{R}}\rangle e^{i\mathbf{k}\mathbf{R}}$ is helpful here and we get the simple scalar equation

$$\frac{\partial}{\partial t}\delta\tilde{n}(\mathbf{k}, t) = -\tilde{\Gamma}(\mathbf{k})\delta\tilde{n}(\mathbf{k}, t)$$

with

$$\tilde{\Gamma}(\mathbf{k}) = \tilde{w}(0) - \tilde{w}(\mathbf{k}).$$

From the reality of Γ it follows $\tilde{\Gamma}(\mathbf{k}) = \tilde{\Gamma}(-\mathbf{k})^*$, while from the detailed balance follows that $\tilde{\Gamma}(\mathbf{k})$ is real. On the other hand, the inequality

$$\sum_{\mathbf{R}} w(\mathbf{R})e^{i\mathbf{k}\mathbf{R}} \leq \sum_{\mathbf{R}} |w(\mathbf{R})e^{i\mathbf{k}\mathbf{R}}|$$

implies $\tilde{\Gamma}(\mathbf{k}) > 0$ for $\mathbf{k} \neq 0$. The total particle number conservation in Fourier transforms looks as $\delta\tilde{n}(0) = 0$. The assumed finite range of the transition rates implies, that one may develop $\tilde{\Gamma}(\mathbf{k})$ around $\mathbf{k} = 0$ and the lowest term is a quadratic one $\tilde{\Gamma}(\mathbf{k}) \approx Dk^2$, with

$$D = \frac{1}{6}\sum_{\mathbf{R}} w(\mathbf{R})\mathbf{R}^2 > 0.$$

Asymptotically for $t \to \infty$ the behavior of the particle density is dominated by the small k values of the Fourier transform for which

$$\delta\tilde{n}(\mathbf{k}, \mathbf{t}) \approx e^{-Dk^2 t}\delta\tilde{n}(\mathbf{k}, 0)$$

and this corresponds exactly to the behavior predicted by the diffusion equation

$$\frac{\partial}{\partial t}n(\mathbf{x}, t) = -D\nabla^2 n(\mathbf{x}, t).$$

Therefore, we may identify the constant D with the diffusion constant of this hopping model.

5.4.5 Transverse Magneto-Resistance in Ultra-Strong Magnetic Field

As we have earlier discussed in Sect. 2.2.2, the Landau states of an electron in a constant magnetic field are characterized by a discrete quantum number n, a wave vector k_z in the field direction and the coordinate X of the center of cyclotron motion in the plane perpendicular to the magnetic field. The extension of the wave-function in the x direction is determined by the magnetic length $\ell_B = \sqrt{\frac{\hbar c}{|e|B}}$, therefore in a very strong magnetic field $\mathbf{B}$ one might expect, that at least along the x axis the state is strongly localized, and the rapidly, with the frequency $\omega_c = \frac{|e|B}{mc}$, oscillating relative coordinate is irrelevant for transport phenomena. In this sense, one might conceive the transport in the x direction as a hopping problem in continuum from one X to another one X' due to the interaction with phonons or lattice defects.

In the presence of an electric field $\mathbf{E} = (E, 0, 0)$ in the x direction the degeneracy of the Landau state is lifted. Intuitively one may expect a potential energy contribution

$$\varepsilon_{n, X, k_z,} = \frac{\hbar^2 k_z^2}{2m} + \hbar\omega_c (n + \frac{1}{2}) + eEX \qquad (n = 0, 1, 2, ...) .$$

(More precisely, one may show that there is also a constant non-relevant energy shift $-\frac{e^2 E^2}{m\omega_c}$, that we may ignore.)

However, if hopping between states of different coordinates X is is allowed, the contribution of the field produced by the non-homogeneous distribution of the electron charge modifies in time the effective on-site field and finally a self-consistent in-homogeneous equilibrium distribution arises which screens the electric field inside far away from the surface. We shall not discuss here this kind of evolution, but consider another situation in which we admit that a steady flow state with constant electric field is achieved by attaching some external electron sources at the ends of the system along the x axis.

The average X coordinate is

$$\langle X \rangle \equiv \sum_\nu X \langle n_\nu \rangle$$

with the simplifying notation $\nu \equiv n, k_z, X$.

The average velocity by hopping (omitting the contribution of the contacts on the boundaries!) is

$$\langle \dot{X} \rangle = \sum_\nu X \frac{d}{dt} \langle n_\nu \rangle = - \sum_{\nu, \nu'} X \{ W_{\nu\nu'} \langle n_\nu \rangle (1 - \langle n_{\nu'} \rangle) - W_{\nu'\nu} \langle n_{\nu'} \rangle (1 - \langle n_\nu \rangle) \} ,$$

where the transition rates due to phonons or random potentials (impurities) satisfy the detailed balance

$$W_{\nu\nu'} = W_{\nu'\nu} e^{\beta(\varepsilon_\nu - \varepsilon_{\nu'})} .$$

We assume now, that a steady flow state occurs in which the charge density is unchanged i.e. the average occupation numbers are the same as in equilibrium in the absence of the electric field

$$\langle n_{n,k_z,X} \rangle = f(\varepsilon_{n,k_z}^0); \qquad \varepsilon_{n,k_z}^0 = \frac{\hbar^2 k_z^2}{2m} + \hbar\omega_c(n + \frac{1}{2})$$

with $f(\varepsilon)$ being the Fermi function. (Later we shall check the consistency of this assumption.)

Introducing a symmetrical matrix $\tilde{W}_{\nu\nu'}$ by

$$W_{\nu\nu'} = \tilde{W}_{\nu\nu'} e^{\frac{\beta}{2}(E_\nu - E_{\nu'})}; \qquad \tilde{W}_{\nu\nu'} = \tilde{W}_{\nu'\nu}$$

we may then write

$$\langle \dot{X} \rangle = -\sum_{\nu,\nu'} X \tilde{W}_{\nu\nu'} \left\{ e^{\frac{\beta}{2}(E_\nu - E_{\nu'})} f(\varepsilon_{n,k_z}^0)\left(1 - f(\varepsilon_{n',k_z'}^0)\right) \right.$$
$$\left. - e^{-\frac{\beta}{2}(E_\nu - E_{\nu'})} f(\varepsilon_{n',k_z'}^0)\left(1 - f(\varepsilon_{n,k_z}^0)\right) \right\} .$$

Now, if one is interested only in the electric conductivity one may retain only the first order terms in the electric field. Since the terms inside the big bracket are vanishing in the absence of the field we may replace $\tilde{W}_{\nu\nu'}$ by its equilibrium value $w_{\alpha,X;\alpha',X'}$ in the absence of the electric field (here we introduced the notation α for the subset of quantum numbers n, k_z) and retain only the term of first order in the field from the exponentials in the big bracket

$$\langle \dot{X} \rangle \approx -e\beta E \sum_{\alpha,X} \sum_{\alpha',X'} X(X - X') w_{\alpha,X;\alpha',X'} f(\varepsilon_{n,k_z}^0)\left(1 - f(\varepsilon_{n',k_z'}^0)\right)$$
$$= -\frac{1}{2}e\beta E \sum_{\alpha,X} \sum_{\alpha',X'} (X - X')^2 w_{\alpha,X;\alpha',X'} f(\varepsilon_\alpha^0)\left(1 - f(\varepsilon_{\alpha'}^0)\right) .$$

Then we get the current density along the x axis as

$$j_x \equiv \frac{e}{\Omega}\langle \dot{X} \rangle = -\frac{1}{2\Omega}e^2\beta E \sum_{\alpha,X}\sum_{\alpha',X'}(X - X')^2 w_{\alpha,X;\alpha',X'} f(\varepsilon_\alpha^0)\left(1 - f(\varepsilon_{\alpha'}^0)\right)$$

and we may identify the transverse conductivity (Titeica formula)

$$\sigma_{xx} = \frac{1}{2\Omega}e^2\beta E \sum_{\alpha,X}\sum_{\alpha',X'}(X - X')^2 w_{\alpha,X;\alpha',X'} f(\varepsilon_\alpha^0)\left(1 - f(\varepsilon_{\alpha'}^0)\right)$$

where Ω is the volume of the system. Of course, as usual the thermodynamic limit has still to be performed.

Let us now check the validity of our assumption about the stationary flow solution. That means $f(\varepsilon_\alpha^0)$ should be a stationary solution of the rate equation, at least to first order in the electric field. Under the same kind of expansion, we get the requirement

$$\sum_{\alpha',X'} (X - X')w_{\alpha,X;\alpha',X'} f(\varepsilon_\alpha^0) \left(1 - f(\varepsilon_{\alpha'}^0)\right) = 0$$

If the transition rates depend only on the distance $|X - X'|$ and fall rapidly with it, then far away from the boundaries this condition is obviously satisfied. At the boundaries one might suppose the presence of some external sources to feed the current flow.

Chapter 6
Optical Properties

The linear response to a time dependent adiabatic external perturbation as well as the equilibrium linear response are presented. The specific relationships of the longitudinal dielectric response of an electron-hole plasma to the density-density correlation function depending on the explicit inclusion of the Coulomb interaction between the charged particles is explained. The dielectric function is explicitly calculated within the Hartree approximation. The response to a transverse (propagating) electric field within a simple model of a semiconductor with s-type conduction and p-type valence bands including the Coulomb attraction between the created electron-hole pair leads to the Elliot-formula containing the exciton peak, as well as the Coulomb-enhancement. Some discussion is devoted also to a modern branch of optical experiments with ultra-short and intense laser beams. The so-called semiconductor Bloch equations are derived and a third order non-linear approach to differential transmission (DTS) and four wave mixing (FWM) is developed.

6.1 Linear Response to a Time Dependent External Perturbation

In this Chapter we shall discuss the interaction of quantum mechanical charged particle systems with external electromagnetic fields in optical domain (light). The treatment first will be restricted to phenomena in the linear domain and we shall develop the conventional formalism for this purpose. Later we shall treat also some topics of nonlinear optics related to the interaction of matter with intense laser pulses.

© Springer International Publishing AG, part of Springer Nature 2018   93
L. A. Bányai, *A Compendium of Solid State Theory*,
https://doi.org/10.1007/978-3-319-78613-1_6

Let us consider a quantum-mechanical system defined by the Hamilton operator H. We want to look for the response of this system to a weak time-dependent external perturbation $H'(t)$. The total Hamiltonian we consider is

$$H_{tot}(t) = H + H'(t).$$

The state of a quantum mechanical statistical ensemble is described by the density operator $R(t)$ and the average $\langle A \rangle$ of an observable A is given by $Tr\{R(t)A\}$. The density operator is hermitian, positive definite $R(t) \geq 0$ (its eigenvalues are non-negative) and is normalized $Tr\{R(t)\} = 1$. The time evolution is governed by the Liouville equation

$$i\hbar \frac{\partial R(t)}{\partial t} = [H_{tot}(t), R(t)].$$

In the interaction picture defined by the transformation

$$\bar{R}(t) = e^{\frac{iH(t-t_0)}{\hbar}} R(t) e^{-\frac{iH(t-t_0)}{\hbar}}$$

the Liouville equation looks as

$$i\hbar \frac{\partial \bar{R}}{\partial t} = [\bar{H}'(t), \bar{R}] \quad ; \qquad \bar{H}'(t) = e^{\frac{iH(t-t_0)}{\hbar}} H'(t) e^{-\frac{iH(t-t_0)}{\hbar}}.$$

Let us consider that at some time t_0 before the introduction of the perturbation the system was in thermal equilibrium

$$R(t_0) = R_0 \equiv \frac{e^{-\beta(H-\mu N)}}{Tr\{e^{-\beta(H-\mu N)}\}}.$$

Then for the deviation from the equilibrium

$$\bar{R} = R_0 + \delta\bar{R}$$

we have

$$i\hbar \frac{\partial \delta\bar{R}(t)}{\partial t} = [\bar{H}'(t), R_0] + \dots .$$

where the unspecified terms must be of higher order in the perturbation. The formal solution with the initial condition

$$\delta\bar{R}(t_0) = 0$$

is.

$$\delta\bar{\rho}(t) = \frac{1}{i\hbar} \int_{t_0}^{t} dt' [\bar{H}'(t'), \rho_0] + \dots$$

or

$$R(t) = R_0 + \frac{1}{\imath \hbar} \int_{t_0}^t dt' e^{\frac{\imath H(t'-t)}{\hbar}} \left[H'(t'), R_0 \right] e^{\frac{-\imath H(t'-t)}{\hbar}} + \ldots$$

We are interested here only in terms linear in the perturbation and consequently, omit all higher order terms. Then the average of an observable A is given by

$$\langle A(t) \rangle - \langle A \rangle_0 = \frac{1}{\imath \hbar} \int_{t_0}^t dt' Tr \left\{ e^{\frac{\imath H(t'-t)}{\hbar}} \left[H'(t'), R_0 \right] e^{\frac{-\imath H(t'-t)}{\hbar}} A \right\}$$

$$= \frac{1}{\imath \hbar} \int_{t_0}^t dt' \langle \left[A_H(t - t'), H'(t') \right] \rangle_0$$

where we used cyclical permutability under the trace and $\langle \ldots \rangle_0$ denotes the equilibrium average. This defines the linear response of the system to a time dependent perturbation.

6.2 Equilibrium Linear Response

There is another kind of linear response problem, that we describe now. Let us suppose, that the system is in equilibrium in the presence of a time-independent external perturbation. An example of such a situation we considered earlier was the penetration of a static electric field into a semiconductor. Here we treat the general case in order to underline the difference to the previous time-dependent non-equilibrium problem.

Thus, we consider a system described by a time-independent Hamiltonian, containing a small perturbation H'

$$H_{tot} = H + H'$$

in a state described by the macro-canonical equilibrium density operator

$$R = \frac{e^{-\beta(H_{tot} - \mu N)}}{Tr \left\{ e^{-\beta(H_{tot} - \mu N)} \right\}} .$$

We want to find out the deviation of the equilibrium average of a certain observable A in the presence of the perturbation from its equilibrium average in the absence of the perturbation

$$\delta \langle A \rangle = \langle A \rangle - \langle A \rangle_0 = Tr \left\{ RA \right\} - Tr \left\{ R_0 A \right\} ,$$

where

$$R = \frac{e^{-\beta(H_{tot}-\mu N)}}{Tr\left\{e^{-\beta(H_{tot}-\mu N)}\right\}}; \quad R_0 = \frac{e^{-\beta(H-\mu N)}}{Tr\left\{e^{-\beta(H-\mu N)}\right\}}.$$

We develop the exponential operator in powers of the perturbation H' using

$$e^{-\beta(H+H'-\mu N)} = e^{-\beta(H-\mu N)}\left\{1 - \int_0^\beta d\lambda e^{\lambda(H-\mu N)} H' e^{-\lambda(H-\mu N)} + \cdots\right\}$$

to get the equilibrium linear response

$$\delta\langle A\rangle = \int_0^\beta d\lambda \langle\left[H'(-\imath\hbar\lambda) - \langle H'\rangle_0\right](A - \langle A\rangle_0)\rangle_0$$

There are two important remarks to be made here:

(i) We admitted implicitly that the chemical potential μ and the inverse tempera-
     ture β remain unchanged by the introduction of the perturbation, however this
     implies, that the average energy and the average number of particles is not the
     same. Alternatively, we could have fixed the average particle and energy den-
     sities and include the corresponding variation of the chemical potential and
     temperature.
(ii) It is not at all obvious, that the zero-frequency limit of the adiabatic time depen-
     dent linear response coincides with the equilibrium linear response result.

The origin of this discrepancy lies in the delicate problem of irreversibility. Only
irreversible processes can bring a system to equilibrium and these are not included
in the quantum-mechanical evolution. See the previous discussion of irreversibility
in Chap. 5.

6.3 Dielectric Response of a Coulomb Interacting Electron Plasma

Let us now consider the perturbation of a Coulomb interacting electron plasma (with a
uniform compensating positive background) by an external longitudinal electric field
defined by an external potential $V^{ext}(\mathbf{x}, t)$. In equilibrium the average charge density
$< \rho(\mathbf{x}) >_0$ is supposed to be vanishing everywhere, while due to the perturbation

$$H'(t) = \int d\mathbf{x}\rho(\mathbf{x})V^{ext}(\mathbf{x}, t)$$

it will be different from zero. By pushing $t_0 \to -\infty$, the linear response to this
perturbation is

$$\langle\rho(\mathbf{x}, t)\rangle = \int_{-\infty}^{\infty} dt' \int d\mathbf{x}' \kappa(\mathbf{x}, \mathbf{x}'; t - t') V^{ext}(\mathbf{x}', t')$$

with the kernel being the charge density correlation function

$$\kappa(\mathbf{x}, \mathbf{x}'; t) \equiv \frac{\theta(t)}{\imath\hbar} \langle[\rho(\mathbf{x}, t)_H , \rho(\mathbf{x}', 0)_H]\rangle_0 .$$

In our homogeneous system, this function may depend only on the difference of the coordinates $\kappa(\mathbf{x}, \mathbf{x}'; t) \equiv \kappa(\mathbf{x} - \mathbf{x}'; t)$. While extending the integration in time to $-\infty$, one has to consider the introduction of the perturbation in an adiabatic manner

$$V^{ext}(\mathbf{x}, t) = e^{\imath(\omega+\imath 0)t} V^{ext}(\mathbf{x}) ,$$

in order to have well defined Fourier transforms. The symbol $+\imath 0$ here means, that one considers a small positive adiabatic parameter that after the performed integration goes to zero.

After a Fourier transformation in the time and space variables we get

$$\langle\tilde{\rho}(\mathbf{k}, \omega)\rangle = \tilde{\kappa}(\mathbf{k}, \omega)\tilde{V}^{ext}(\mathbf{k}, \omega)$$

with

$$\tilde{\kappa}(\mathbf{k}, \omega) = \int_0^{\infty} dt \int d\mathbf{x} e^{-\imath\mathbf{kx}} e^{\imath(\omega+\imath 0)t} \kappa(\mathbf{x}, t).$$

Now, it is very important to realize, that in a Coulomb system the local charge density at its turn is the source of an internal potential. We are interested in the relation between the internal charge density and the total potential $V = V^{ext} + V^{int}$ that is defined by the Poisson equation (here in Fourier transforms)

$$k^2 \tilde{V}(\mathbf{k}, \omega) = 4\pi \langle\tilde{\rho}(\mathbf{k}, \omega)\rangle + 4\pi \rho^{ext}(\mathbf{k}, \omega)$$

while

$$k^2 \tilde{V}^{ext}(\mathbf{k}, \omega) = 4\pi \rho^{ext}(\mathbf{k}, \omega).$$

From these equations and the linear response of the charge density to the external potential we get

$$\varepsilon(\mathbf{k}, \omega)k^2 \tilde{V}(\mathbf{k}, \omega) = 4\pi \rho^{ext}(\mathbf{k}, \omega)$$

with the dielectric function

$$\varepsilon(\mathbf{k}, \omega) \equiv \frac{1}{1 + \frac{4\pi}{k^2}\tilde{\kappa}(\mathbf{k}, \omega)} .$$

The situation is quite different if one treats the Coulomb interacting electron plasma in the s.c. Hartree approximation. Then the perturbation is the self-consistent potential itself and not the external one. Therefore, from the Poisson equation one gets

$$\varepsilon(\mathbf{k}, \omega)^{sc} \equiv 1 - \frac{4\pi}{k^2} \tilde{\kappa}^{sc}(\mathbf{k}, \omega) .$$

However, the two density correlation functions would be quite different!

6.4 Dielectric Function of an Electron Plasma in the Hartree Approximation

As we have seen before, in the self-consistent Hartree approximation one has free electrons with modified s.c. one-particle energies. In what follows we ignore this modification and therefore our electrons in the absence of the electric field are treated just to be free. The charge density operator in terms of the second quantized wave functions is

$$\rho(\mathbf{x}) = e\psi(\mathbf{x})^+\psi(\mathbf{x}) .$$

Here we ignored the positive uniform compensating charge, that plays no role in the following discussion. In terms of the creation annihilation operators of the one-particle states of wave-vector $\mathbf{k}$, in the Heisenberg picture

$$\rho(\mathbf{r}, t)_H = \frac{e}{\Omega} \sum_{\mathbf{k}} \sum_{\mathbf{k'}} c_{\mathbf{k}}^+ c_{\mathbf{k'}} e^{\imath(\mathbf{k'}-\mathbf{k})\mathbf{r}} e^{\frac{\imath}{\hbar}(e_{\mathbf{k}}-e_{\mathbf{k'}})t}$$

with the kinetic energies $e_{\mathbf{k}} = \frac{\hbar^2 k^2}{2m}$. The density correlation function is

$$\kappa^{sc}(\mathbf{r}, t) = \frac{e^2}{\imath\hbar\Omega^2} \sum_{\mathbf{k}} \sum_{\mathbf{k'}} \sum_{\mathbf{p}} \sum_{\mathbf{p'}} e^{\frac{\imath}{\hbar}(e_{\mathbf{k}}-e_{\mathbf{k'}})t} e^{\imath(\mathbf{k'}-\mathbf{k})\mathbf{r}} \langle [c_{\mathbf{k}}^+ c_{\mathbf{k'}}, c_{\mathbf{p}}^+ c_{\mathbf{p'}}] \rangle_0$$

The average $\langle [c_{\mathbf{k}}^+ c_{\mathbf{k'}}, c_{\mathbf{p}}^+ c_{\mathbf{p'}}] \rangle_0$ may be easily performed by knowing the commutator

$$[c_{\mathbf{k}}^+ c_{\mathbf{k'}}, c_{\mathbf{p}}^+ c_{\mathbf{p'}}] = \delta_{\mathbf{k'}\mathbf{p}} c_{\mathbf{k}}^+ c_{\mathbf{p'}} - \delta_{\mathbf{p'}\mathbf{k}} c_{\mathbf{p}}^+ c_{\mathbf{k'}}$$

and retaining the only surviving average

$$\langle c_{\mathbf{k}}^+ c_{\mathbf{p'}} \rangle_0 = \delta_{\mathbf{k}\mathbf{p'}} f_{\mathbf{k}} ,$$

with $f_{\mathbf{k}} \equiv\, < c_{\mathbf{k}}^+ c_{\mathbf{k}} >_0$ being the Fermi function. One gets finally

$$\varepsilon_L(\mathbf{q}, \omega) = 1 - \frac{4\pi e^2}{\Omega q^2} \sum_{\mathbf{k}} \frac{f_{\mathbf{k}+\mathbf{q}} - f_{\mathbf{k}}}{e_{\mathbf{k}+\mathbf{q}} - e_{\mathbf{k}} - \hbar\omega - \imath 0} .$$

In the evaluation of this formula one has to take into account the identity

$$\frac{1}{x - \iota 0} = P\left(\frac{1}{x}\right) + \iota \pi \delta(x)$$

with the first term being the principal value, while the second one contains Dirac's delta function. Then

$$\Re \varepsilon_L(\mathbf{q}, \omega) = 1 - \frac{4\pi e^2}{\Omega q^2} \sum_{\mathbf{k}} \frac{2 f_{\mathbf{k}}(e_{\mathbf{k}} - e_{\mathbf{k+q}})}{(e_{\mathbf{k}} - e_{\mathbf{k+q}})^2 + (\hbar\omega)^2}$$

and

$$\Im \varepsilon_L(\mathbf{q}, \omega) = \frac{4\pi^2 e^2}{\Omega q^2} \sum_{\mathbf{k}} \left(f_{\mathbf{k}} - f_{\mathbf{k+q}}\right) \delta(\hbar\omega + e_{\mathbf{k}} - e_{\mathbf{k+q}})$$

$$= \frac{4\pi^2 e^2}{\Omega q^2} \sum_{\mathbf{k}} \left(f(e_{\mathbf{k}}) - f(e_{\mathbf{k}} + \hbar\omega)\right) \delta(\hbar\omega + e_{\mathbf{k}} - e_{\mathbf{k+q}})$$

$$= \frac{e^2 m^2}{\hbar^4 q^3} \int_0^\infty de\, (f(e) - f(e + \hbar\omega))\, \theta \left(1 - \left|\frac{\hbar\omega - \frac{\hbar^2 q^2}{2m}}{2\sqrt{e\frac{\hbar^2 q^2}{2m}}}\right|\right)$$

At $T = 0^0 K$ we get

$$\Im \varepsilon_L(\mathbf{q}, \omega) = \frac{e^2 m^2}{\hbar^4 q^3} \int_{E_F - \hbar\omega}^{E_F} de\theta \left(1 - \left|\frac{\hbar\omega - \frac{\hbar^2 q^2}{2m}}{2\sqrt{e\frac{\hbar^2 q^2}{2m}}}\right|\right).$$

and the imaginary part of the dielectric function vanishes for

$$\left(\hbar\omega - \frac{\hbar^2 q^2}{2m}\right)^2 > \frac{2\hbar^2 q^2}{m} E_F.$$

At any temperature asymptotically for $\omega \to \infty$

$$\Im \varepsilon_L(\mathbf{q}, \omega) \approx 0$$

and

$$\Re \varepsilon_L(\mathbf{q}, \omega) \approx 1 - \frac{4\pi e^2 \langle n \rangle}{m\omega^2} = 1 - \left(\frac{\omega_{pl}}{\omega}\right)^2$$

and

$$\omega_{pl} = \sqrt{\frac{4\pi e^2 \langle n \rangle}{m}}$$

is the plasma frequency. Since the dielectric constant vanishes for big $\omega = \omega_{pl}$ (provided ω_{pl} is also big !), eigenoscillations of the plasma may occur in the absence of an external field at this frequency.

Peculiar properties has also the static longitudinal dielectric function

$$\varepsilon_L(\mathbf{q}, 0) = 1 - \frac{4\pi e^2}{\Omega q^2} \sum_{\mathbf{k}} P \left(\frac{f_{\mathbf{k+q}} - f_{\mathbf{k}}}{e_{\mathbf{k+q}} - e_{\mathbf{k}}} \right) .$$

For $\mathbf{q} \to 0$

$$\varepsilon_L(\mathbf{q}, 0) \approx 1 - \frac{4\pi e^2}{\Omega q^2} \sum_{\mathbf{k}} \frac{\partial f(e_{\mathbf{k}})}{\partial e_{\mathbf{k}}} = 1 - \frac{4\pi e^2}{q^2} \int dez(e) \frac{\partial f(e)}{\partial e} = 1 + \frac{\kappa^2}{q^2}$$

with

$$\kappa^2 \equiv 4\pi e^2 \int dez(e) \frac{\partial f(e)}{\partial e} \approx \begin{cases} \frac{4\pi}{k_B T} \langle n \rangle & (T \to \infty) \\ \frac{6\pi e^2}{E_F} \langle n \rangle & (T \to 0) \end{cases}$$

This last result shows, that if one puts an external point-like charge q inside the plasma ($V^{ext}(\mathbf{r}) = \frac{q}{r}$ i.e. $\tilde{V}^{ext}(\mathbf{q}) = \frac{4\pi}{q^2}$), then the resulting potential $\tilde{V}(q)$ for small q looks as

$$\tilde{V}(\mathbf{q}) = \frac{\tilde{V}^{ext}(\mathbf{q})}{\varepsilon_L(\mathbf{q}, 0)} \approx \frac{4\pi e^2}{q^2 + \kappa^2}$$

and in real space at big distances in the real space it will look as a Yukawa potential

$$V(\mathbf{r}) \approx \frac{e^2}{|\mathbf{r}|} e^{-\kappa r} .$$

It is important to remark, that in the whole discussion here no interaction with any thermal bath was considered and therefore there were no irreversible effects incorporated.

6.5   The Transverse, Inter-band Dielectric Response of an Electron-Hole Plasma

A transverse propagating electric field $\mathbf{E}(\mathbf{x}, t)$ (photon) may be defined by the choice of a vanishing scalar potential ($V(\mathbf{x}, t) = 0$) and a transverse vector potential ($\nabla \mathbf{A}(\mathbf{x}, t) = 0$). The interaction of charged particles with this electromagnetic field (ignoring the $\mathbf{A}^2$ term) is given by

$$-\int d\mathbf{x} \mathbf{j}(\mathbf{x}) \mathbf{A}(\mathbf{x}, t) .$$

In a direct gap semiconductor, due to its vector character, this interaction may couple mainly s-like states of the conduction band with p-like states of the degenerate valence band.

In an electron-hole plasma model therefore one describes such an electromagnetic interaction with a dipole Hamiltonian

$$H_{em}(t) = d \sum_{\mu=1}^{3} \sum_{\sigma=\pm\frac{1}{2}} \int d\mathbf{x} \psi_{e,-\sigma}(\mathbf{x})^+ \psi_{h,\mu,\sigma}(\mathbf{x})^+ \mathcal{E}_\mu(\mathbf{x}, t) + h.c. \ .$$

The second quantized wave functions of the holes $\psi_{h,\mu,\sigma}(\mathbf{x})$ correspond to the three p-type valence bands of spin σ. Here d is a phenomenological dipole constant and the spin is $\sigma = \pm\frac{1}{2}$. The above Hamiltonian containing pairs of creation and annihilation operators indicates, that a photon having an energy $\hbar\omega$ above the band gap may create an electron-hole pair.

We want to take into account here, that the newly created electron and hole interact through Coulomb forces. As we shall see, this fact has important consequences for the absorption spectrum. The Coulomb interaction we shall treat within the intra-band s.c. Fock approximation by retaining the non-local average (at instant t)

$$\langle \psi_{e,\sigma}(\mathbf{x}) \psi_{h,\mu,-\sigma}(\mathbf{x}') \rangle_t \ .$$

Such an average vanishes in the absence of the electric field, but appears after the introduction of the field.

Within this approximation one has an s.c. Coulomb energy

$$\sum_{\mu=1}^{3} \sum_{\sigma=\pm\frac{1}{2}} \int d\mathbf{x} \int d\mathbf{x}' \frac{e^2}{|\mathbf{x}-\mathbf{x}'|} \left(\psi_{e,\sigma}(\mathbf{x}) \psi_{h,\mu,-\sigma}(\mathbf{x}') \langle \psi_{e,\sigma}(\mathbf{x}) \psi_{h,\mu,-\sigma}(\mathbf{x}') \rangle_t^* + h.c. \right) \ .$$

Obviously, this term has the same operator structure as $H_{em}(t)$ and is at least of first order in the field. Therefore, the effective perturbation consists of $H_{em}(t)$ plus an induced perturbation of the same structure

$$H'(t) = d \sum_{\mu=1}^{3} \int d\mathbf{x} \mathcal{P}_\mu(\mathbf{x}, \mathbf{x}) \mathcal{E}_\mu(\mathbf{x}, t)$$

$$+ \sum_{\mu=1}^{3} \int d\mathbf{x} \int d\mathbf{x}' \frac{e^2}{|\mathbf{x}-\mathbf{x}'|} \mathcal{P}_\mu(\mathbf{x}, \mathbf{x}') \langle \mathcal{P}_\mu(\mathbf{x}, \mathbf{x}') \rangle_t^* + h.c.$$

where the notation

$$\mathcal{P}_\mu(\mathbf{x}, \mathbf{x}') \equiv \sum_{\sigma=\pm\frac{1}{2}} \psi_{e,\sigma}(\mathbf{x}) \psi_{h,\mu,-\sigma}(\mathbf{x}')$$

was introduced.

Let us consider now the Heisenberg equations of motion for the operator $\mathscr{P}_\mu(\mathbf{x}, \mathbf{x}')$. Using the commutation relation

$$\left[\psi_{e,\sigma}(\mathbf{x})\psi_{h_\mu,-\sigma}(\mathbf{x}'), \psi_{h_\nu,-\sigma}(\mathbf{y}')^+\psi_{e,\sigma}(\mathbf{y})^+\right] = \delta_{\mu\nu}\delta(\mathbf{x}-\mathbf{y})\delta(\mathbf{x}'-\mathbf{y}')$$
$$-\delta(\mathbf{x}-\mathbf{y})\psi_{h,-\sigma}(\mathbf{y}')^+\psi_{h,-\sigma}(\mathbf{x}') - \delta(\mathbf{x}'-\mathbf{y}')\delta_{\mu\nu}\psi_{e,\sigma}(\mathbf{y})^+\psi_{e,\sigma}(\mathbf{x})$$

we get

$$\left\{\imath\hbar\frac{\partial}{\partial t} + \frac{\hbar^2}{2m_e}\nabla_e^2 + \frac{\hbar^2}{2m_h}\nabla_h^2 + \frac{e^2}{|\mathbf{x}_e - \mathbf{x}_h|}\right\}\mathscr{P}_\mu(\mathbf{x}_e, \mathbf{x}_h) =$$

$$d\sum_{\nu=1}^{3}\sum_\sigma\int d\mathbf{y}\,\mathscr{E}_\nu(\mathbf{y}, t)\left(\delta_{\mu\nu}\delta(\mathbf{x}_e - \mathbf{y})\delta(\mathbf{x}_h - \mathbf{y})\right.$$

$$\left.-\delta_{\mu\nu}\delta(\mathbf{x}_h - \mathbf{y})\psi_{e,\sigma}(\mathbf{x}_h)^+\psi_{e,\sigma}(\mathbf{x}_e) - \delta(\mathbf{x}_e - \mathbf{y})\psi_{h_\mu,-\sigma}(\mathbf{x}_e)^+\psi_{h_\nu,-\sigma}(\mathbf{x}_h)\right)$$

$$-\sum_{\nu=1}^{3}\sum_\sigma\int d\mathbf{x}\int d\mathbf{x}'\frac{e^2}{|\mathbf{x}-\mathbf{x}'|}\langle\mathscr{P}_\nu(\mathbf{x}, \mathbf{x}')\rangle_t\left(\delta_{\mu\nu}\delta(\mathbf{x} - \mathbf{x}_e)\delta(\mathbf{x}' - \mathbf{x}_h)\right. =$$

$$-\delta_{\mu\nu}\delta(\mathbf{x}_h - \mathbf{x}')\psi_{e,\sigma}(\mathbf{x})^+\psi_{e,\sigma}(\mathbf{x}_e) - \delta(\mathbf{x}_e - \mathbf{x})\psi_{h_\mu,-\sigma}(\mathbf{x}')^+\psi_{h_\nu,-\sigma}(\mathbf{x}_h)\right) .$$

Now, we perform the average $\langle\ldots\rangle_t$ by taking into account, that in the absence of the field there were no electrons and holes. The initial state is the electron-hole vacuum and even thereafter the population appears only in second order in the field. Therefore we may consider

$$\langle\psi_{\alpha,\sigma}(\mathbf{x})^+\psi_{\alpha,\sigma}(\mathbf{x}')\rangle_t \approx 0; \quad (\alpha = e, h)$$

In the linear approximation in the external field we get therefore the equation of motion

$$\left\{\imath\hbar\frac{\partial}{\partial t} + \frac{\hbar^2}{2m_e}\nabla_e^2 + \frac{\hbar^2}{2m_h}\nabla_h^2 + \frac{e^2}{|\mathbf{x}_e - \mathbf{x}_h|} + E_g\right\}\langle\mathscr{P}_\mu(\mathbf{x}_e, \mathbf{x}_h)\rangle_t = d\mathscr{E}_\mu(\mathbf{x}_e, t)\delta(\mathbf{x}_e - \mathbf{x}_h)$$

where E_g is the band gap energy.

The solution of this non-homogeneous equation one gets by using the Green function

$$G(\mathbf{x}_e, \mathbf{x}_h; \mathbf{x}'_e, \mathbf{x}'_h; t)$$

of the electron-hole Schroedinger equation

$$\left\{\imath\hbar\frac{\partial}{\partial t} + \frac{\hbar^2}{2m_e}\nabla_e^2 + \frac{\hbar^2}{2m_h}\nabla_h^2 + \frac{e^2}{|\mathbf{x}_e - \mathbf{x}_h|}\right\}G(\mathbf{x}_e, \mathbf{x}_h; \mathbf{x}'_e, \mathbf{x}'_h; t) =$$

$$\delta(\mathbf{x}_e - \mathbf{x}'_e)\delta(\mathbf{x}_h - \mathbf{x}'_h)\delta(t) .$$

On his turn it may be built up from the eigenfunctions $\Phi_\alpha(\mathbf{x}_e, \mathbf{x}_h)$ of the Coulomb interacting electron-hole Schroedinger equation

$$\left\{ -\frac{\hbar^2}{2m_e}\nabla_e^2 - \frac{\hbar^2}{2m_h}\nabla_h^2 - \frac{e^2}{|\mathbf{x}_e - \mathbf{x}_h|} \right\} \Phi_\alpha(\mathbf{x}_e, \mathbf{x}_h) = \varepsilon_\alpha \Phi_\alpha(\mathbf{x}_e, \mathbf{x}_h)$$

as

$$G_r(\mathbf{x}_e, \mathbf{x}_h; \mathbf{x}'_e, \mathbf{x}'_h; t) = \frac{1}{\imath\hbar}\theta(t) \sum_\alpha \Phi_\alpha(\mathbf{x}_e, \mathbf{x}_h)_\alpha \Phi_\alpha(\mathbf{x}'_e, \mathbf{x}'_h)^* e^{\imath\frac{\varepsilon_\alpha}{\hbar}t} \ .$$

Here we considered the retarded Green function relevant for our physical problem. One may separate the center of mass motion

$$\mathbf{X} \equiv \frac{m_e \mathbf{x}_e + m_h \mathbf{x}_h}{m_e + m}$$

from the relative motion and

$$\Phi_{\mathbf{q},l,m,\varepsilon}(\mathbf{x}_e, \mathbf{x}_h) = \frac{1}{\sqrt{\Omega}} e^{-\imath\mathbf{q}\mathbf{X}} \phi_{l,m,\varepsilon}(\mathbf{x}) \ ,$$

where $\phi_{l,m,\varepsilon}(\mathbf{x})$ are the well-known Coulomb wave functions characterized by their angular momentum l, m and energy ε while Ω is the volume of the system. These states may belong to the discrete spectrum or to the continuum.

Using the Green function $G(\mathbf{x}_e, \mathbf{x}_h; \mathbf{x}'_e, \mathbf{x}'_h; t)$ one finds

$$\langle \mathscr{P}_\mu(\mathbf{x}_e, \mathbf{x}_h) \rangle_t = \int_{-\infty}^t dt' \int d\mathbf{x}'_e \int d\mathbf{x}'_h e^{\frac{\imath}{\hbar}E_g(t-t')} G(\mathbf{x}_e, \mathbf{x}_h; \mathbf{x}'_e, \mathbf{x}'_h; t - t') d\mathscr{E}_\mu(\mathbf{x}'_e, t')\delta(\mathbf{x}'_e - \mathbf{x}'_h)$$

or

$$\langle \mathscr{P}_\mu(\mathbf{x}, \mathbf{X}) \rangle_t = \frac{d}{\imath\hbar\Omega} \sum_{\mathbf{q},l,m,\varepsilon} \int_{-\infty}^t dt' e^{\frac{\imath}{\hbar}(\varepsilon + E_g + \frac{\hbar^2 q^2}{2(m_e+m_h)})(t-t')}$$

$$\times \int d\mathbf{X}' e^{-\imath\mathbf{q}(\mathbf{X}-\mathbf{X}')} \phi_{l,m,\varepsilon}(\mathbf{x})\phi_{l,m,\varepsilon}(0)^* \mathscr{E}_\mu(\mathbf{X}', t') \ .$$

The transverse, inter-band dielectric function is defined by the relation of the inter-band polarization $\mathbf{P}(\mathbf{x}, t) = d\langle \mathscr{P}(0, \mathbf{x}) \rangle_t$ to the electric field

$$\mathbf{P}(\mathbf{q}, \omega) = \frac{1}{4\pi}(\varepsilon_T(\mathbf{q}, \omega) - 1)\mathbf{E}_t(\mathbf{q}, \omega)$$

and we get

$$\varepsilon_T(\mathbf{q}, \omega) = 1 - 4\pi d^2 \sum_\varepsilon |\phi_{0,0,\varepsilon}(0)|^2$$

$$\times \left(\frac{1}{\hbar\omega - E_g - \varepsilon - \frac{\hbar^2 q^2}{2(m_e+m_h)} + \iota 0} - \frac{1}{\hbar\omega + E_g + \varepsilon + \frac{\hbar^2 q^2}{2(m_e+m_h)} + \iota 0} \right).$$

Here we took into account that in $\mathbf{x} = 0$ only the s-wave functions contribute.

The imaginary part, that describes the absorption spectrum is given then explicitly by the Elliott -formula

$$\Im\varepsilon_T(\mathbf{q}, \omega) = \frac{4d^2}{a_B^3 E_R} \left[\sum_{n=1}^\infty \frac{1}{n^3}\delta\left(\Delta + \frac{1}{n^2}\right) + \frac{1}{2}\theta(\Delta)\frac{1}{1 - e^{-\frac{2\pi}{\sqrt{\Delta}}}} \right]$$

with

$$\Delta \equiv \frac{\hbar\omega - E_g - \frac{\hbar^2 q^2}{2(m_e+m_h)}}{E_R},$$

where E_R is the excitonic Rydberg energy.

In the absence of the Coulomb interaction ($e = 0$) we have had the simple inter-band absorption spectrum

$$\Im\varepsilon_T(\mathbf{q}, \omega)^0 = \frac{d^2}{\pi a_B^3 E_R}\theta(\Delta)\sqrt{\Delta}.$$

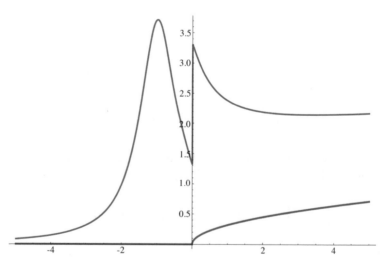

Fig. 6.1 Absorption spectra as function of the detuning Δ at the band edge ($\Delta = 0$) with Coulomb (blue) and without Coulomb effects (red)

In order to represent graphically in Fig. 6.1 the expected absorption spectrum we had to introduce by hand an element of irreversibility through an arbitrary line-width for the discrete exciton lines of which we retained only the two lowest states, but only the ground state may be actually seen. The Coulomb interaction between the electron and the hole created by photon absorption introduces the exciton resonance as well as a Coulomb enhancement of the continuum at the threshold as to be compared with the red line of absorption without Coulomb effects. See also the comparison with the exciton state density Fig. 3.1 calculated within the tight binding Wannier scheme.

6.6   Ultra-Short-Time Spectroscopy of Semiconductors

In the last decades interesting experiments were performed with very intense and ultra-short (on femtosecond scale!), laser beams on semiconductors. Their purpose is to obtain information about ultra-short time processes. These experiments are theoretically treated by the extension of the linear response including higher order terms in the development. However, drastic simplifications are needed. First of all, we shall ignore the coordinate dependence of the electric field $\mathbf{E}$ (negligible photon momentum!) and consider

$$\mathbf{E}(t) = \mathscr{E}(t) \cos(\omega_0 t) \,,$$

where ω_0 is the carrier frequency in the optical domain and $\mathscr{E}(t)$ is the pulse envelope. This may be justified by the fact, that for photon energy in the neighborhood of the energy gap the photon wave length λ in the most interesting semiconductors is much bigger than the exciton radius a_B.

The system we consider is again an electron-hole plasma as in Sect. 6.5 with an s-like conduction band and a p-like valence band. The phenomenological coupling to the transverse electromagnetic field one uses is that of an electric dipole in analogy to the similar atomic coupling. We chose the quantization direction of the valence band states so, that only one of the p bands couples to the field and therefore we may ignore the vector notation.

$$H_{em}(t) = \hbar \omega_R(t) \cos(\omega_0 t) \sum_\sigma \int d\mathbf{x} \left(\psi_{e,\sigma}(\mathbf{x})^+ \psi_{h,-\sigma}(\mathbf{x})^+ + h.c. \right)$$

$$= \hbar \omega_R(t) \cos(\omega_0 t) \sum_{\sigma,\mathbf{k}} \left(a_{e,\sigma,\mathbf{k}} a_{h,-\sigma,-\mathbf{k}} + h.c. \right) \,,$$

where $\omega_R(t) = \frac{1}{\hbar} d\mathscr{E}(t)$ is the so called (here, time dependent!) Rabi frequency.

Obviously, higher order terms in the development of the system response would produce also oscillations with multiples of the frequency ω_0. Such terms may be less important if already $2\hbar\omega_0$ is far away from the band gap, and we are eager to discard them. An elegant way to do this is to ignore some of the terms in the perturbation and consider the "rotating wave" Hamiltonian

$$H_{em}^{(rw)} = \frac{1}{2}\hbar\omega_R(t)e^{\iota\omega_0 t}\sum_{k,\sigma}a_{e,\sigma,k}a_{h,-\sigma,-k} + h.c. \; .$$

The total Hamiltonian we start with will be

$$H(t) = H_{eh}^{HF}(t) + H_{em}^{(rw)} \; ,$$

where the electron-hole part is taken in the Hartree Fock approximation

$$
\begin{aligned}
H_{eh}^{HF}(t) = &\sum_{\sigma}\sum_{k}\left(\left(\frac{\hbar^2 k^2}{2m_e} + \frac{1}{2}E_g\right)a_{e,\sigma,k}^{+}a_{e,\sigma,k} + \left(\frac{\hbar^2 k^2}{2m_h} + \frac{1}{2}E_g\right)a_{h,\sigma,k}^{+}a_{h,\sigma,k}\right) \\
&+ \sum_{\sigma,\sigma'}\sum_{k,k',q}\frac{4\pi e^2}{q^2 V}\Big[a_{e,\sigma,k+q}^{+}a_{e,\sigma,k}\langle a_{e,\sigma',k'-q}^{+}a_{e,\sigma',k'}\rangle_t \\
&- a_{e,\sigma,k+q}^{+}a_{e,\sigma'k'}\langle a_{e,\sigma',k'-q}^{+}a_{e,\sigma,k}\rangle_t \\
&+ a_{h,\sigma,k+q}^{+}a_{h,\sigma k}\langle a_{h,\sigma',k'-q}^{+}a_{h,\sigma',k'}\rangle_t \\
&- a_{h,\sigma,k+q}^{+}a_{h,\sigma'k'}\langle a_{h,\sigma',k'-q}^{+}a_{h,\sigma,k}\rangle_t \\
&- a_{e,\sigma,k+q}^{+}a_{e,\sigma k}\langle a_{h,\sigma',k'-q}^{+}a_{h,\sigma',k'}\rangle_t \\
&- a_{h,\sigma,k+q}^{+}a_{h,\sigma k}\langle a_{e,\sigma',k'-q}^{+}a_{e,\sigma',k'}\rangle \\
&- a_{e,\sigma,k+q}^{+}a_{h,\sigma',k'-q}^{+}\langle a_{h,\sigma',k'}a_{e,\sigma,k}\rangle_t \\
&- a_{h,\sigma,k'}a_{e,\sigma',k}\langle a_{e,\sigma',k+q}^{+}a_{h,\sigma,k'-q}^{+}\rangle_t\Big] \; .
\end{aligned}
$$

Here we took into account all the anomalous averages that will be induced only by the electromagnetic field.

We may take into account also momentum conservation, spin independence as well as the charge neutrality of the total system. Then

$$\langle a_{e,\sigma',k'}^{+}a_{e,\sigma,k}\rangle_t = \delta_{k,k'}\delta_{\sigma,\sigma'}f_{e,k}(t)$$

$$\langle a_{h,\sigma',k'}^{+}a_{h,\sigma,k}\rangle_t = \delta_{k,k'}\delta_{\sigma,\sigma'}f_{h,k}(t)$$

$$\langle a_{e,\sigma',k'}a_{h,\sigma,k}\rangle_t = \delta_{k,-k'}\delta_{\sigma,-\sigma'}p_k(t)$$

$$\sum_{k}f_{e,k}(t) = \sum_{k}f_{h,k}(t)$$

and the Hartree terms disappear. Then the following equations of motion emerge for the electron-hole populations $f_{e,k}(t)$, $f_{h,k}(t)$ and the "inter-band polarization" p_k

$$\frac{\partial}{\partial t} f_{e,\mathbf{k}}(t) = \Im \left\{ \Omega_\mathbf{k}^*(t)\, p_\mathbf{k}(t) e^{\imath \omega_0 t} \right\}$$

$$\frac{\partial}{\partial t} f_{h,\mathbf{k}}(t) = \Im \left\{ \Omega_\mathbf{k}^*(t)\, p_\mathbf{k}(t) e^{\imath \omega_0 t} \right\}$$

$$\frac{\partial}{\partial t} p_\mathbf{k}(t) + \frac{\imath}{\hbar} \left(\varepsilon_{e,\mathbf{k}}(t) + \varepsilon_{h,\mathbf{k}}(t) \right) p_\mathbf{k}(t) = \frac{\imath}{2} \Omega_k(t) e^{-\imath \omega_0 t} \left(1 - f_{e,\mathbf{k}}(t) - f_{h,\mathbf{k}}(t) \right) \quad .$$

Here $\varepsilon_{e,\mathbf{k}}(t)$ and $\varepsilon_{h,\mathbf{k}}(t)$ are the"renormalized" electron and hole energies

$$\varepsilon_{e,\mathbf{k}}(t) \equiv \frac{\hbar^2 k^2}{2m_e} + \frac{1}{2} E_g + \frac{1}{V} \sum_{\mathbf{k}' \neq \mathbf{k}} V_{|\mathbf{k}-\mathbf{k}'|}\, f_{e,\mathbf{k}'}(t)$$

$$\varepsilon_{h,\mathbf{k}}(t) \equiv \frac{\hbar^2 k^2}{2m_h} + \frac{1}{2} E_g + \frac{1}{V} \sum_{\mathbf{k}' \neq \mathbf{k}} V_{|\mathbf{k}-\mathbf{k}'|}\, f_{h,\mathbf{k}'}(t)$$

while $\Omega_\mathbf{k}(t)$ is the generalized Rabi-frequency

$$\Omega_\mathbf{k}(t) \equiv \omega_R(t) + \frac{2}{\hbar} \frac{1}{V} \sum_{\mathbf{k}' \neq \mathbf{k}} V_{|\mathbf{k}-\mathbf{k}'|}\, p_{\mathbf{k}'}(t) e^{\imath \omega_0 t} \; ; \quad V_q \equiv \frac{4\pi e^2}{q^2} \quad .$$

The rapidly oscillating factor $e^{-\imath \omega_0 t}$ may be eliminated from the equation by a redefinition of the inter-band polarization

$$p_\mathbf{k}(t) \equiv \bar{p}_\mathbf{k}(t) e^{-\imath \omega_0 t}$$

and we obtain the "Semiconductor Bloch Equations" for the slowly varying entities $f_{e,\mathbf{k}}, f_{h,\mathbf{k}}, \bar{p}_\mathbf{k}$

$$\frac{\partial}{\partial t} f_{e,\mathbf{k}} = \Im \left\{ \Omega_\mathbf{k}^* \, \bar{p}_\mathbf{k} \right\}$$

$$\frac{\partial}{\partial t} f_{h,\mathbf{k}} = -\Im \left\{ \Omega_\mathbf{k}^* \, \bar{p}_\mathbf{k} \right\}$$

$$\left(\frac{\partial}{\partial t} + \frac{\imath}{\hbar} \Delta_\mathbf{k} \right) \bar{p}_\mathbf{k} = \frac{\imath}{2} \Omega_k \left(1 - f_{e,\mathbf{k}} - f_{h,\mathbf{k}} \right) - \frac{\bar{p}_\mathbf{k}}{T_2} \quad .$$

with the detuning

$$\Delta_\mathbf{k} \equiv \varepsilon_{e,\mathbf{k}} + \varepsilon_{h,\mathbf{k}} - \hbar \omega_0 \, .$$

We have introduced here by hand also a phenomenological term with a "relaxation time" T_2 in order to take into account symbolically other interactions. This is a standard approach in the treatment of the non-linear optics of two level atoms in the frame of the Bloch Equations on a quite different time scale. In semiconductors, the chosen time scale is such, that the here ignored processes are still coherent and only on a longer time scale they destroy the coherence. The ultra-short-time experiments are performed precisely to study the kinetics of the electron-hole plasma on this time scale, where coherent quantum-mechanical features play still an essential role. The many-body treatment, of these rapid processes, taking into account Coulomb interactions as well as the interaction with LO-phonons is called "quantum kinetics". This theory is rather complicated and we do not touch it.

Since the populations of electrons and holes in the absence of scattering are equal $f_e = f_h \equiv f$, one of these equations is superfluous. The above equations may be solved either numerically or by a systematic development of the solution in powers of the field i.e. of $\omega_R(t)$.

6.7   Third Order Non-linear Response

We outline here the scheme of the third order nonlinear response. Since different $\mathbf{k}$ values in the Semiconductor Bloch Equations are not coupled we may omit the $\mathbf{k}$ index. Further, we simplify the discussion by omitting here the Hartree-Fock terms. The solution we look for is a systematic development in the powers of the electromagnetic perturbation

$$f(t) = f^{(1)}(t) + f^{(2)}(t) + f^{(3)}(t) \dots$$
$$\bar{p}(t) = \bar{p}^{(1)}(t) + \bar{p}^{(2)}(t) + \bar{p}^{(3)}(t) + \dots \quad .$$

To first order we have

$$f^{(1)}(t) = 0$$
$$\bar{p}^{(1)}(t) = \frac{\imath}{2} \int_{-\infty}^{t} dt' \, e^{-(\frac{\imath}{\hbar}\Delta + \frac{1}{T_2})(t-t')} \omega_R(t') \quad ,$$

while to third order

$$f^{(3)}(t) = 0$$
$$\bar{p}^{(3)}(t) = -\imath \int_{-\infty}^{t} dt' \, e^{-(\frac{\imath}{\hbar}\Delta + \frac{1}{T_2})(t-t')} \omega_R(t') f^{(2)}(t') \quad .$$

Inserting in the last step, the result of the previous approximation we get the only non-vanishing terms

$$f^{(2)}(t) = \frac{1}{2} \int_{-\infty}^{t} dt' \int_{-\infty}^{t'} dt'' \, \Re \left(\omega_R^*(t') \omega_R(t'') \, e^{-(\frac{\imath}{\hbar}\Delta + \frac{1}{T_2})(t'-t'')} \right)$$

and

$$\bar{p}^{(3)}(t) = -\frac{\iota}{2} \int_{-\infty}^{t} dt' \int_{-\infty}^{t'} dt'' \int_{-\infty}^{t''} dt''' e^{-(\frac{\iota}{\hbar}\Delta+\frac{1}{T_2})(t-t')} \omega_R(t')$$
$$\times \Re\left(\omega_R^*(t'') \omega_R(t''') e^{-(\frac{\iota}{\hbar}\Delta+\frac{1}{T_2})(t''-t''')}\right) .$$

In what follows we shall consider that the optical field (laser) consists of a stronger pump pulse and a weaker, much shorter test-pulse, both having the same carrier frequency ω_0, also

$$\omega_R(t) = \omega_R^P(t) + \omega_R^T(t) .$$

Since the test-pulse is assumed to be very weak, we will retain only terms of first order in $\omega_R^T(t)$. (This restriction is however not compulsory in the analysis of the four-wave-mixing.) Then

$$\bar{p}^{(3)}(t) \approx -\frac{\iota}{2} \int_{-\infty}^{t} dt' \int_{-\infty}^{t'} dt'' \int_{-\infty}^{t''} dt''' e^{-(\frac{\iota}{\hbar}\Delta+\frac{1}{T_2})(t-t')} \omega_R^T(t')$$
$$\times \Re\left(\omega_R^{P*}(t'') \omega_R^P(t''') e^{-(\frac{\iota}{\hbar}\Delta+\frac{1}{T_2})(t''-t''')}\right)$$
$$-\frac{\iota}{2} \int_{-\infty}^{t} dt' \int_{-\infty}^{t'} dt'' \int_{-\infty}^{t''} dt''' e^{-(\frac{\iota}{\hbar}\Delta+\frac{1}{T_2})(t-t')} \omega_R^P(t')$$
$$\times \Re\left[\left(\omega_R^{P*}(t'') \omega_R^T(t''') + \omega_R^{T*}(t'') \omega_R^P(t''')\right) e^{-(\frac{\iota}{\hbar}\Delta+\frac{1}{T_2})(t''-t''')}\right] .$$

As one may see, there are two different structures, either

$$\bar{p} \propto \omega_R^T \cdot |\omega_R^P|^2$$

or

$$\bar{p} \propto \omega_R^{T*} \cdot (\omega_R^P)^2 .$$

To interpret the significance of these structures we must return to the ignored propagation properties of the laser beams given by the phase factors $e^{\iota \mathbf{k}_T \mathbf{r}}$ respectively $e^{\iota \mathbf{k}_P \mathbf{r}}$. While by the interaction within the semiconductor, these may be neglected, now their importance is very important in analyzing non-linear experiments. Since the excited inter-band polarization is itself a source of emerging electromagnetic waves, we expect that the first structure will cause an emerging beam in the $\mathbf{k}_T$ -direction, while the second in the $(2\mathbf{k}_P - \mathbf{k}_T)$-direction. The first is responsible for the differential transmission (DTS), while the second for the four wave mixing (FWM). The configuration of a DTS experiment is shown in Fig. 6.2, while that of an FWM experiment in Fig. 6.3. Here the pink arrows show the test and pump propagation. The pulses themselves are shown in violet, while the new signal in the $(2\mathbf{k}_P - \mathbf{k}_T)$ direction is shown with a red arrow.

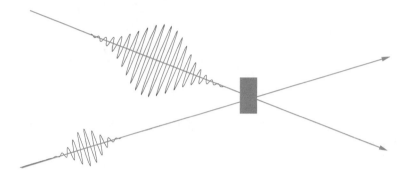

Fig. 6.2 Differential transmission experiment

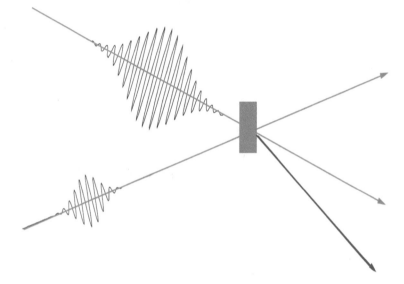

Fig. 6.3 Four-wave-mixing experiment

6.8 Differential Transmission

In this experiment, one uses two laser pulses of different widths with a retardation time τ between them. By the stronger pump pulse electron hole pairs are excited, and the test pulse sees a prepared state, different from that in the absence of the pump. To identify the transmitted test pulse from the pump pulse one chooses a small angle between the pulses. By varying the retardation time τ one may obtain informations about the evolution of the interacting electron-hole pairs created by the pump pulse.

Within the χ_3 theory outlined before we have

$$\bar{p}_{\mathbf{k}}^{(3)DTS}(t) = -\frac{i}{4} \int_{-\infty}^{t} dt' \int_{-\infty}^{t'} dt'' \int_{-\infty}^{t''} dt''' \, e^{-(\frac{i}{\hbar}\Delta_{\mathbf{k}}+\frac{1}{T_2})(t-t')} \, \omega_R^T(t')$$

$$\times \left(\omega_R^{P*}(t'') \omega_R^P(t''') e^{(-\frac{i}{\hbar}\Delta_{\mathbf{k}}-\frac{1}{T_2})(t''-t''')} \right.$$

$$\left. + \omega_R^P(t'') \omega_R^{P*}(t''') e^{(\frac{i}{\hbar}\Delta_{\mathbf{k}}-\frac{1}{T_2})(t''-t''')} \right)$$

$$- \frac{i}{4} \int_{-\infty}^{t} dt' \int_{-\infty}^{t'} dt'' \int_{-\infty}^{t''} dt''' \, e^{-(\frac{i}{\hbar}\Delta_{\mathbf{k}}+\frac{1}{T_2})(t-t')} \, \omega_P(t')$$

$$\times \left[\omega_R^{P*}(t'') \omega_R^T(t''') e^{-(\frac{i}{\hbar}\Delta_{\mathbf{k}}+\frac{1}{T_2})(t''-t''')} \right] \quad .$$

If the pulses do not overlap one may retain only the term

$$\bar{p}^{(3)DTS}(t) \approx -\frac{i}{8} \int_{-\infty}^{t} dt' \, e^{-(\frac{i}{\hbar}\Delta+\frac{1}{T_2})(t-t')} \, \omega_R^T(t') \left| \int_{-\infty}^{t'} dt'' \omega_R^P(t'') e^{\frac{i}{\hbar}\Delta t''} \right|^2 \quad .$$

If the test pulse is very short, one gets further

$$\bar{p}^{(3)DTS}(t) \approx -\frac{i}{8} e^{-(\frac{i}{\hbar}\Delta_{\mathbf{k}}+\frac{1}{T_2})(t-\tau)} \theta(t-\tau) \int_{-\infty}^{\infty} dt' \omega_R^T(t') \left| \int_{-\infty}^{\tau} dt'' \omega_R^P(t'') e^{\frac{i}{\hbar}\Delta_{\mathbf{k}} t''} \right|^2$$

or

$$\bar{p}^{(3)DTS}(t) \approx -\frac{i}{8} \theta(t-\tau) e^{-(\frac{i}{\hbar}\Delta+\frac{1}{T_2})(t-\tau)} \tilde{\omega}_R^T(0) |\tilde{\omega}_R^P(\Delta)|^2 \quad ,$$

where $\tilde{\omega}_R^P(\Delta)$ and $\tilde{\omega}_R^T(0)$ are the Fourier transformed pulse and test Rabi frequencies at the detuning Δ, respectively at $t = 0$.

The total DTS polarization $P^{(3)DTS}(t)$ results after summation over all the wave vectors and re-multiplication with $e^{-i\omega_0 t}$. The Fourier analysis of the resulting time dependent signal shows the modification of the test pulse absorption due to the presence of the pump pulse. This DTS signal (differential transmission) is defined as the difference between the absorption of the test pulse in the presence of the pump-pulse and that in the absence of the pump. Now, it may be shown further that this difference reaches its maximal negative value again at the carrier frequency ω_0, if the test pump retardation time τ is much bigger than the width of the pulses. The meaning of this result is, that one cannot excite electron-hole pairs in a domain, that already was emptied by the pump. This phenomenon is called "hole burning". However, during the time τ the electron-hole system, due to the Coulomb interaction as well as the interaction with phonons, the electron-hole system changes its state. By measuring the DTS signal at different retardations τ one may get information about this evolution and compare it with theoretical predictions. Since in this simple description we did not include such an evolution, we do not follow further the explicit calculus.

6.9   Four Wave Mixing

The FWM signal within the χ_3 theory is contained in the terms proportional to $\omega_R^{T*}(\omega_R^P)^2$

$$
\bar{p}^{(3)VWM}(t) = -\frac{i}{4} \int_{-\infty}^{t} dt' \int_{-\infty}^{t'} dt'' \int_{-\infty}^{t''} dt''' \, e^{-(\frac{i}{\hbar}\Delta+\frac{1}{T_2})(t-t')} \, \omega_R^P(t')
$$
$$
\times \left(\omega_R^{T*}(t'')\,\omega_R^P(t''')\, e^{-(\frac{i}{\hbar}\Delta+\frac{1}{T_2})(t''-t''')} \right.
$$
$$
\left. + \, \omega_R^P(t'')\,\omega_R^{T*}(t''')\, e^{(\frac{i}{\hbar}\Delta-\frac{1}{T_2})(t''-t''')} \right) .
$$

Unlike in the case of the DTS experiment, one chooses here two pulses with a large retardation time τ. Under this condition only the second term of the previous equation contributes

$$
\bar{p}^{(3)VWM}(t) = -\frac{i}{4} \int_{-\infty}^{t} dt' \int_{-\infty}^{t'} dt'' \int_{-\infty}^{t''} dt'''
$$
$$
\times \, e^{\frac{i}{\hbar}\Delta(t'-t+t''-t''')} e^{-\frac{(t-t'+t''-t''')}{T_2}} \, \omega_R^P(t')\omega_R^P(t'')\omega_R^T(t''')
$$

If one considers, that the two pulses do not overlap at all, one gets

$$
P^{(3)VWM}(t) = -\frac{i}{8} e^{-\frac{t}{T_2}} \tilde{\omega}_R^P(0)^2 \tilde{\omega}_R^T(0) \int d\mathbf{k} e^{\frac{i}{\hbar}\Delta_\mathbf{k}(t-2\tau)} .
$$

The integral over $\mathbf{k}$ diverges and one must restrict it to the Brillouin zone. Due to the oscillating factor, the integral reaches its maximum at $t = 2\tau$ and therefore one may observe a "photon echo" in the $2\mathbf{k}_P - \mathbf{k}_T$-direction after a delay time 2τ, its height decreasing as $e^{-\frac{2\tau}{T_2}}$.

The outlined theory of these experiments is oversimplified, illustrating only the essential idea. The deviations from the simple one-particle treatment due to the Coulomb interaction and phonon emission/absorption are the main informations one follows, and the height of the photon echo is not given by the above simple exponential decay. Actually, this kind of optical experiments (due to the retardation time parameter) deliver a sort of motion picture of the evolution of the electron-hole plasma.

Chapter 7
Phase Transitions

A few examples of phase transitions allow an insight into the fine aspects of the thermodynamic limit. Spontaneous symmetry breaking at critical temperatures or densities leads to unexpected stable states characterized by order parameters. We describe at the beginning the ferromagnetic transition within the frame of the Heisenberg model of localized spins. The mathematical subtlety of the Bose-Einstein condensation is discussed in some more details. We devote more place to superconductivity by describing first the phenomenological theory of London, followed by a quantum-mechanical treatment in real space, within the frame of a simple model of an effective interaction between the electrons suggested by the BCS theory.

An important aspect of the physics of condensed matter is the possibility of phase transitions. Their theoretical treatment is essentially based on the thermodynamic limit. We do not intend to describe the multitude of the phase transitions, but just want to take a glimpse at the basic ideas within some specific models. Some of the most spectacular new phenomena although subject of abundant literature are still not ripe for a simple presentation.

7.1 The Heisenberg Model of Ferro-Magnetism

Ferro-magnetism is thought to originate from the spin-dependent exchange interaction of the core electrons of neighboring ions. A simple model is due to Heisenberg that considers N interacting spins $\mathbf{S}$ on a periodic lattice $\mathbf{R}$. The Hamiltonian in the presence of a magnetic field $\mathbf{B} = (0, 0, B)$ oriented along the z axis is

© Springer International Publishing AG, part of Springer Nature 2018

L. A. Bányai, *A Compendium of Solid State Theory*,

https://doi.org/10.1007/978-3-319-78613-1_7

$$H = -\frac{1}{2} \sum_{\mathbf{R}\mathbf{R}'} J(\mathbf{R} - \mathbf{R}')\mathbf{S}_{\mathbf{R}}\mathbf{S}_{\mathbf{R}'} - 2\mu_0 B \sum_{\mathbf{R}} S_{\mathbf{R}}^z \;,$$

where μ_0 is the magneton and the spin-operators $\mathbf{S}$ obey the commutation rules

$$\left[S_{\mathbf{R}}^+, S_{\mathbf{R}}^-\right] = 2S_{\mathbf{R}}^z$$
$$\left[S_{\mathbf{R}}^z, S_{\mathbf{R}}^\pm\right] = \pm S_{\mathbf{R}}^z$$

with $S_{\mathbf{R}}^\pm = S_{\mathbf{R}}^x \pm S_{\mathbf{R}}^y$. The commutators of spin-operators on different sites vanish.

The coupling $J(\mathbf{R}) \geq 0$ is supposed to vanish rapidly with the distance R. One may write the Hamiltonian also in the form

$$H = -\frac{1}{2} \sum_{\mathbf{R}, \mathbf{R}'} J(\mathbf{R} - \mathbf{R}') \left(S_{\mathbf{R}}^z S_{\mathbf{R}'}^z + \frac{S_{\mathbf{R}}^+ S_{\mathbf{R}'}^- + S_{\mathbf{R}}^- S_{\mathbf{R}'}^+}{2} \right) - 2\mu_0 B \sum_{\mathbf{R}} S_{\mathbf{R}}^z$$

to realize, that the total spin $\mathbf{S}_{tot}$ in the z-direction

$$S_{tot}^z \equiv \sum_{\mathbf{R}} S_{\mathbf{R}}^z$$

is conserved

$$\left[S_{tot}^z, H\right] = 0.$$

Therefore, the energy eigenstates are also eigenstates of S_{tot}^z, as well as of $\mathbf{S}_{tot}^2$ due to rotation invariance. A simple basis in the spin space may be built up from products of eigenstates of each spin $|S, m>$ with m being its projection on the z axis.

The state with the maximum of total spin is the one with all the spins aligned in the z-direction

$$|\Phi_0> = |S, S>_{\mathbf{R}_1} |S, S>_{\mathbf{R}_2} \ldots |S, S>_{\mathbf{R}_N}$$

This state is an eigenstate of H

$$H|\Phi_0> = \left(-\frac{1}{2} S^2 \sum_{\mathbf{R}, \mathbf{R}'} J(\mathbf{R} - \mathbf{R}') - 2\mu_0 B N S \right) |\Phi>$$

since

$$S_{\mathbf{R}}^+ |S, S>_{\mathbf{R}} = 0\,.$$

Consider now the average of the Hamiltonian over an arbitrary state $|\Phi>$ of our basis

$$|\Phi> = |S, m_1>_1 |S, m_2>_2 \ldots |S, m_N>_N \;.$$

Since the terms non-commuting with $S_{\mathbf{R}}^z$ vanish in the average, one gets

$$< \Phi|H|\Phi >= -\frac{1}{2} \sum_{R, R'} J(R - R')m_R m_{R'} - 2\mu_0 B \sum_R m_R$$

and this is obviously greater than the eigenenergy of the state $|\Phi_0 >$. Therefore, $|\Phi_0 >$ is the ground state of the system with the ground state energy

$$E_0 = -\frac{1}{2}S^2 \sum_{R, R'} J(R - R') - 2\mu_0 BNS .$$

In this state the total spin in the z direction is NS and the per node magnetic moment (magnetization) is

$$M = 2\mu_0 S .$$

In the absence of the magnetic field B the choice of the quantization axis is arbitrary, and this state would have had a $2NS + 1$ fold degeneracy. The average over all these states of the same energy would give rise to no magnetization. Thus, with a finite number of spins the Heisenberg model would not lead to a stable ferromagnetic state in the absence of an external magnetic field. However, the situation changes drastically if one considers the thermodynamic limit. Indeed, one sees that by $N, \Omega \to \infty$ by $\frac{N}{\Omega} = n$ (constant spin density) even an infinitesimal magnetic field may determine the direction of the magnetization, while its magnitude is the same $M = 2\mu_0 S$. This is the *spontaneous symmetry breaking*.

Let us treat the thermodynamics of this model in the absence of a magnetic field **B** in the mean field (self-consistent) approximation in which one of the spin operators we replace by its average

$$H_{mf} = -\sum_R 2\mu_0 \langle B \rangle S_R$$

with the effective (s.c.) magnetic field being

$$\langle B \rangle = \frac{1}{2\mu_0} \sum_R J(R)\langle S_R \rangle$$

and we look for a solution with the average on site spin $\langle S_R \rangle$ as being site independent and oriented along the z direction. Then

$$H_{mf} = -\sum_R 2\mu_0 \mathcal{B} S_R^z; \qquad \mathcal{B} = \frac{1}{2\mu_0} \langle S^z \rangle \sum_R J(R) .$$

The statistical sum is

$$Z = Spe^{-\beta H_{mf}} = \left(\sum_{S^z} e^{-2\beta\mu_0 \mathcal{B} S^z} \right)^N ,$$

while the free energy and the equilibrium magnetization are given by

$$F \equiv \frac{1}{N} \ln \left(\sum_{S^z} e^{-2\beta\mu_0 \mathcal{B} S^z} \right)$$

and

$$M \equiv -\frac{\partial F}{\partial \mathcal{B}} = 2\beta\mu_0 \langle S^z \rangle$$

For spin $S = \frac{1}{2}$ one gets

$$M = \beta\mu_0 \tanh(\beta\mu_0 \mathcal{B})$$

that has to be considered together with the self-consistency equation

$$\mathcal{B} = \frac{1}{4\mu_0} \tanh(\beta\mu_0 \mathcal{B}) \sum_{\mathbf{R}\neq 0} J(\mathbf{R}) .$$

Since $\frac{\tanh(x)}{x} < 1$, over the critical temperature

$$T_c = \frac{1}{4k_B} \sum_{\mathbf{R}\neq 0} J(\mathbf{R})$$

one has only a vanishing solution for the effective magnetic field $\mathcal{B}$.

However for $T < T_c$ there is a non-vanishing solution and consequently, a finite magnetization in the absence of any external magnetic field. At $T = T_c$ a ferromagnetic phase transition occurs.

7.2 Bose Condensation

Another good example of a phase transition is the Bose condensation. In condensed matter, such a transition may occur for the excitons as massive bosons. Let us consider a system of free massive bosons in a finite volume (with periodical boundary conditions) described by the second quantized Hamiltonian

$$H = \sum_{\mathbf{k}} e_{\mathbf{k}} a_{\mathbf{k}}^+ a_{\mathbf{k}}$$

having the one-particle energies

$$e_{\mathbf{k}} = \frac{\hbar^2 \mathbf{k}^2}{2m} .$$

The macro-canonical statistical distribution gives the average number of bosons in the state of wave-vector $\mathbf{k}$ to be the Bose distribution

$$\frac{1}{e^{\beta(e_k - \mu)} - 1} .$$

Obviously, the chemical potential must be negative $\mu < 0$, otherwise the expression is negative below $k = 0$. The average total number of bosons fixes μ

$$\langle n \rangle = \frac{\langle N \rangle}{\Omega} = \frac{1}{\Omega} \sum_{\mathbf{k}} \frac{1}{e^{\beta(e_k - \mu)} - 1} ,$$

or in the thermodynamic limit

$$\langle n \rangle = \int \frac{d^3k}{(2\pi)^3} \frac{1}{e^{\beta(e_k - \mu)} - 1} .$$

This integral however reaches its possible maximum value at a critical density n_c for $\mu = 0$

$$n_c = \int \frac{d^3k}{(2\pi)^3} \frac{1}{e^{\beta e_k} - 1} = \frac{1}{\sqrt{2\pi^2}} \Gamma(\frac{3}{2}) \zeta(\frac{3}{2}) \left(\frac{m}{\beta \hbar^2} \right)^{\frac{3}{2}} .$$

Seemingly the system cannot accommodate any higher average density! As one may see, the problem lies by the application of the often used rule

$$\frac{1}{\Omega} \sum_{\mathbf{k}'} \rightarrow \int \frac{d^3k}{(2\pi)^3} .$$

Indeed, one may show, that at the critical density n_c, as the volume Ω increases the chemical potential goes to zero from negative values as $\frac{1}{\Omega}$

$$\mu \approx -\frac{const.}{\Omega}$$

and therefore, the contribution of the $\mathbf{k} = 0$ state to the sum is not infinitesimal

$$\frac{1}{\Omega} \frac{1}{e^{-\beta\mu} - 1} \approx \beta const.$$

This contribution has to be added independently. Therefore, the correct equation for $n \geq n_c$ is

$$n_{cond} + \int \frac{d^3k}{(2\pi)^3} \frac{1}{e^{\beta e_k} - 1} = n .$$

Above the critical density the chemical potential remains zero but the condensate density n_{cond} increases.

The critical situation may be achieved either by increasing the density at a fixed temperature ($n \to n_c$, at fixed β), or by lowering the temperature at a fixed density ($\beta \to \beta_c$ at fixed density n).

Like the ferromagnetic phase transition, the Bose condensation shows a spontaneous symmetry breaking. This may be seen within the Bogolyubov approach. In this approach, one introduces an infinitesimal symmetry breaking term in the Hamiltonian before the thermodynamic limit and let it vanish after the limit is already performed.

Bogolyubov introduces terms that break invariance against a multiplication of the creation/annihilation operators by a phase factor, therefore particle number conservation is violated. In the macro-canonical statistical sum

$$Z = Tr\{e^{-\beta(H-\mu N)}\}$$

one considers

$$H - \mu N = \sum_{\mathbf{k}}(e_{\mathbf{k}} - \mu)a_{\mathbf{k}}^{+}a_{\mathbf{k}} + \lambda^{*}\sqrt{\Omega}a_{0} + \lambda\sqrt{\Omega}a_{0}^{+}$$

with a small parameter λ. A simple c-number shift

$$A_{0} = a_{0} - \frac{\lambda\sqrt{\Omega}}{\mu}$$

defines the new bosonic annihilation operators for the state $\mathbf{k} = 0$ and helps to bring the expression again to a quadratic form

$$H - \mu N = \sum_{\mathbf{k}\neq 0}(e_{\mathbf{k}} - \mu)a_{\mathbf{k}}^{+}a_{\mathbf{k}} - \mu A_{0}^{+}A_{0} + \frac{|\lambda|^{2}\Omega}{\mu} \ .$$

The free energy is then

$$F = \frac{1}{\beta\Omega}\sum_{\mathbf{k}}\ln\{1 - e^{-\beta(e_{\mathbf{k}}-\mu)}\} + \frac{|\lambda|^{2}}{\mu}$$

leading to the average boson density

$$\langle n \rangle = \frac{\langle N \rangle}{\Omega} = \frac{1}{\Omega}\sum_{\mathbf{k}}\frac{1}{e^{\beta(e_{\mathbf{k}}-\mu)} - 1} + \frac{|\lambda|^{2}}{\mu^{2}} \ .$$

Now we may apply without hesitating the limit rule for getting integrals out of sums, since the chemical potential remains always negative and get

$$\langle n \rangle = \frac{|\lambda|^2}{\mu^2} + \int \frac{d^3k}{(2\pi)^3} \frac{1}{e^{\beta(e_k - \mu)} - 1} \ .$$

At the critical density n_c the chemical potential vanishes in the $|\lambda| \to 0$ limit as

$$\mu \to \frac{|\lambda|}{\sqrt{n_{cond}}}$$

and we get the former result.

A new aspect of this approach is that here we explicitly recognize an order parameter λ that is the analog of the magnetization with the surviving phase of the symmetry breaking parameter ($\lambda = |\lambda| e^{\iota\phi}$) and the average value $\langle a_0 \rangle$ of the original annihilation operator of the $\mathbf{k} = 0$ state becomes macroscopic (proportional to the square root of the total number of condensate bosons $N_{cond} = n_{cond}\Omega$)

$$\langle a_0 \rangle = e^{\iota\phi}\sqrt{N_{cond}} \ .$$

A characteristic property of the condensed state is the existence of long range space correlations. The Riemann sum in the correlation function

$$\langle \psi(\mathbf{x})\psi^+(\mathbf{y}) \rangle \equiv \frac{1}{\Omega} \sum_{\mathbf{k}} e^{\iota\mathbf{k}(\mathbf{x}-\mathbf{y})} \langle a_{\mathbf{k}}^+ a_{\mathbf{k}} \rangle = \frac{1}{\Omega} \sum_{\mathbf{k}} e^{\iota\mathbf{k}(\mathbf{x}-\mathbf{y})} \frac{1}{e^{\beta(e_{\mathbf{k}} - \mu)} - 1}$$

above the critical temperature T_c goes over into the integral

$$\int \frac{d^3k}{(2\pi)^3} \frac{e^{\iota\mathbf{k}(\mathbf{x}-\mathbf{y})}}{e^{\beta e_{\mathbf{k}}} - 1}$$

that vanishes as a power law for $|\mathbf{x} - \mathbf{y}| \to \infty$, while below the critical temperature one gets

$$\langle \psi(\mathbf{x})\psi^+(\mathbf{y}) \rangle = n_{cond} + \int \frac{d^3k}{(2\pi)^3} \frac{e^{\iota\mathbf{k}(\mathbf{x}-\mathbf{y})}}{e^{\beta e_{\mathbf{k}}} - 1}$$

and there is a term surviving over infinite distances.

7.3 Superconductivity

A phase transition of special art is superconductivity of some materials below a certain critical temperature. Its peculiarity is, that it reveals not only spectacular equilibrium properties like the Meissner effect (ideal diamagnetism expelling magnetic fields), but also non-usual non-equilibrium properties as the flow of electric current without resistance. Besides, connectivity properties of the material, by allowing circulating permanent currents also play a decisive role. We shall describe in what follows the phenomenological theory of London which succeeds in a simple mathematical formulation to reflect the essentials of the phenomena. Thereafter we discuss a quantum mechanical model in real space, suggested by the BCS (Bardeen, Cooper, Schrieffer) theory of an effective electron-electron attraction mediated by phonons.

7.3.1   The Phenomenological Theory of London

The macroscopic Maxwell equations for the electric and magnetic fields and their sources (charge and current densities) are

$$\nabla \mathbf{B} = 0$$

$$\nabla \times \mathbf{E} = -\frac{1}{c}\frac{\partial}{\partial t}\mathbf{B}$$

$$\nabla \times \mathbf{B} = \frac{4\pi}{c}\mathbf{j} + \frac{1}{c}\frac{\partial}{\partial t}\mathbf{E}$$

$$\nabla \mathbf{E} = 4\pi\rho$$

The charges and currents obviously have to satisfy the conservation rule

$$\nabla\mathbf{j} + \frac{\partial}{\partial t}\rho = 0 \ .$$

These equations alone do not determine the macroscopic electromagnetic fields. They have to be supplemented by "matter relations" for the current- and charge- densities characteristic for the given material, which relates these entities again to the fields.
    The simplest examples are:

- *normal conductors* (metals), with no charge density inside and Ohms law for the current density
$$\mathbf{j}_{Ohm} = \sigma\mathbf{E}, \qquad \rho = 0 \ .$$

- *insulators (dielectrics)*, with polarization charges and polarization currents characterized by a dielectric constant
$$\mathbf{j}_{pol} = \frac{1}{4\pi}(\varepsilon - 1)\frac{\partial}{\partial t}\mathbf{E}, \qquad \rho = -\frac{1}{4\pi}(\varepsilon - 1)\nabla\mathbf{E} \ .$$

- *semiconductors*, which beside the ohmic and polarization currents also have a diffusion current induced by the "free" charge density ρ_{free}
$$\mathbf{j}_{diff} = -D\nabla\rho_{free}.$$

The diffusion constant D is related to the conductivity by the Einstein relation defined by the carrier density n and the temperature

$$\sigma = ne^2 k_B T D .$$

- *magnetic materials* with the specific magnetization current density we do not mention explicitly here.

To describe the basic aspects of a superconductor, London proposed two simple phenomenological "matter relations", that indeed successfully describe the main properties of these materials.

The stipulated equations that relate the current density to the electromagnetic field are characterized by a phenomenological constant Λ

$$\mathbf{E} = \Lambda \frac{\partial}{\partial t}\mathbf{j}$$
$$\Lambda \nabla \times \mathbf{j} = -\frac{1}{c}\mathbf{B}.$$

The first London equation was suggested by the behavior of free electrons, that are accelerated by an electric field. If one takes the curl of this equation and uses the Maxwell equations, one gets

$$\frac{\partial}{\partial t}\left(\Lambda \nabla \times \mathbf{j} + \frac{1}{c}\mathbf{B}\right) = 0 .$$

Therefore, the second London equation states only, that not just the time derivative of the expression under the bracket is vanishing, but the expression itself.

Now, taking the curl of the third Maxwell equation

$$\nabla \times \nabla \times \mathbf{B} = \frac{4\pi}{c}\nabla \times \mathbf{j} + \frac{1}{c}\frac{\partial}{\partial t}\nabla \times \mathbf{E}$$

and using this second London equation, as well as the identity

$$\nabla \times \nabla \times \mathbf{B} = \nabla(\nabla \mathbf{B}) - \nabla^2 \mathbf{B}$$

one gets

$$\nabla^2 \mathbf{B} - \frac{4\pi}{c^2 \Lambda}\mathbf{B} - \frac{1}{c^2}\frac{\partial^2}{\partial t^2}\mathbf{B} = 0 .$$

In a stationary regime with constant fields it looks as

$$\nabla^2 \mathbf{B} - \frac{4\pi}{c^2 \Lambda}\mathbf{B} = 0 .$$

The length: $\ell = c\sqrt{\frac{\Lambda}{4\pi}}$ obviously defines the penetration depth of a magnetic field inside a superconductor in equilibrium (Meissner effect). (Compare with the penetration of a static electric field into a semiconductor.)

After integrating the second London equation on a surface S

$$\int_S d\mathbf{s}\nabla \times \mathbf{j} = -\frac{1}{\Lambda c}\int_S d\mathbf{s}\mathbf{B}$$

and using (Stokes theorem) we get

$$\oint d\mathbf{l} \; \mathbf{j} = -\frac{1}{\Lambda c}\Phi$$

also, a circulation of the current around a loop may be sustained by a magnetic flux inside the loop. Since in equilibrium there is no magnetic field inside the superconductor only the multi-connectivity of the sample allows for permanent circular currents.

These two main properties characterize the equilibrium properties of a superconductor. In what concerns the first London equation it concerns non-equilibrium phenomena. These are much difficult to treat. The microscopical theories of superconductivity concentrate mostly on the equilibrium properties: the existence of a phase transition and the Meissner effect.

7.4 Superconducting Phase Transition in a Simple Model of Electron-Electron Interaction

The fundamental idea for a microscopical explanation of superconductivity stems from Bardeen-Cooper-Schrieffer (BCS) about the effective interaction between the electrons of opposite spins that might become attractive, at least in a limited range of momenta (in the neighborhood of the Fermi energy), due to exchange of optical phonons. This attraction produces no bound states, but a correlation between electrons of opposite spins and momenta, with anomalous electron number non-conserving averages analogously to the former description of the Bose condensation. This BCS-theory offers a correct interpretation of the properties of the superconducting phase transition.

On the other hand, the theory of the electromagnetic properties must be formulated in the real space. Here we want to explore the possibilities to implement the BCS model into a model based on an electron-electron interaction by a fictitious potential without any claim for a realistic theory of superconductivity. We shall treat within this approach also the explanation of the Meissner effect.

We consider instead of the Hamilton operator H, the operator $\mathscr{H} \equiv H - \mu N$, which is more appropriate in tackling problems where only the average number of electrons $\langle N \rangle$ is fixed

$$
\begin{aligned}
\mathscr{H} &\equiv H - \mu N \\
&= \sum_{\sigma = \pm \frac{1}{2}} \int d\mathbf{x} \psi_\sigma^+(\mathbf{x}) \left\{ \frac{1}{2m} \left(-\imath \hbar \nabla - \frac{e}{c} \mathbf{A}(\mathbf{x}) \right)^2 - \mu \right\} \psi_\sigma(\mathbf{x}) \\
&\quad + \frac{1}{2} \int d\mathbf{x} \int d\mathbf{x}' V(\mathbf{x} - \mathbf{x}') \{ \psi_{\frac{1}{2}}^+(\mathbf{x}) \psi_{-\frac{1}{2}}^+(\mathbf{x}') \psi_{\frac{1}{2}}(\mathbf{x}') \psi_{-\frac{1}{2}}(\mathbf{x}) + h.c. \}
\end{aligned}
$$

with some "attractive" potential $V(\mathbf{x})$ in the presence of a constant magnetic field described by the vector potential $\mathbf{A}(\mathbf{x})$ in the Coulomb gauge $\nabla \mathbf{A}(\mathbf{x}) = 0$. The chosen interaction does not lead to bound states, like an attractive Coulomb potential would do, but, as we shall see, it gives rise to BCS—type correlations.

We approximate the problem within the Hartree-Fock-Bogolyubov scheme by retaining only the anomalous (electron number non-conserving) averages with opposite spins

$$
\begin{aligned}
\mathscr{H}_{HFB} &= \sum_\sigma \int d\mathbf{x} \psi_\sigma^+(\mathbf{x}) \left\{ \frac{1}{2m} \left(-\imath \hbar \nabla - \frac{e}{c} \mathbf{A}(\mathbf{x}) \right)^2 - \mu \right\} \psi_\sigma(\mathbf{x}) \\
&\quad + \int d\mathbf{x} \int d\mathbf{x}' V(\mathbf{x} - \mathbf{x}') \left\{ \langle \psi_{\frac{1}{2}}^+(\mathbf{x}) \psi_{-\frac{1}{2}}^+(\mathbf{x}') \rangle \psi_{-\frac{1}{2}}(\mathbf{x}') \psi_{\frac{1}{2}}(\mathbf{x}) \right. \\
&\quad \left. + \langle \psi_{-\frac{1}{2}}(\mathbf{x}') \psi_{\frac{1}{2}}(\mathbf{x}) \rangle \psi_{\frac{1}{2}}^+(\mathbf{x}) \psi_{-\frac{1}{2}}^+(\mathbf{x}') - \langle \psi_{\frac{1}{2}}^+(\mathbf{x}) \psi_{-\frac{1}{2}}^+(\mathbf{x}') \rangle \langle \psi_{-\frac{1}{2}}(\mathbf{x}') \psi_{\frac{1}{2}}(\mathbf{x}) \rangle \right\} .
\end{aligned}
$$

Here the average is defined over the macro-canonical distribution

$$
\langle \ldots \rangle \equiv \frac{Tr \left\{ e^{-\beta \mathscr{H}_{HFB}} \ldots \right\}}{Tr \left\{ e^{-\beta \mathscr{H}_{HFB}} \right\}} .
$$

In the absence of the magnetic field ($\mathbf{A} = 0$), in a finite volume Ω with discrete wave vectors we get the unperturbed "Hamiltonian" in terms of the creation and annihilation operators of the $\mathbf{k}, \sigma$ states

$$
\begin{aligned}
\mathscr{H}^0{}_{HFB} = \sum_\mathbf{k} &\left\{ \varepsilon_\mathbf{k} \left(a_{\mathbf{k}, \frac{1}{2}}^+ a_{\mathbf{k}, \frac{1}{2}} + a_{-\mathbf{k}, -\frac{1}{2}}^+ a_{-\mathbf{k}, -\frac{1}{2}} \right) \right. \\
&\left. + \Delta_\mathbf{k} a_{-\mathbf{k}, -\frac{1}{2}} a_{\mathbf{k}, \frac{1}{2}} + \Delta_\mathbf{k}^* a_{\mathbf{k}, \frac{1}{2}}^+ a_{-\mathbf{k}, -\frac{1}{2}}^+ + \mathscr{C}(\mathbf{k}) \right\}
\end{aligned}
$$

with

$$\varepsilon(\mathbf{k}) = \frac{\hbar^2 k^2}{2m} - \mu \,,$$

$$\Delta_{\mathbf{k}} \equiv \frac{1}{\Omega} \sum_{\mathbf{k}'} V(\mathbf{k} - \mathbf{k}') \langle a^+_{\mathbf{k}',\frac{1}{2}} a^+_{-\mathbf{k}',-\frac{1}{2}} \rangle$$

and

$$\tilde{V}(\mathbf{k}) = \int_{\Omega} dx e^{i\mathbf{k}x} V(\mathbf{x}) \,.$$

This "Hamiltonian" may be diagonalized with the Bogolyubov-Valatin canonical transformation

$$\begin{pmatrix} a_{\mathbf{k},\frac{1}{2}} \\ a^+_{-\mathbf{k},-\frac{1}{2}} \end{pmatrix} = \begin{pmatrix} u_{\mathbf{k}} & v_{\mathbf{k}} \\ -v^*_{\mathbf{k}} & u^*_{\mathbf{k}} \end{pmatrix} \begin{pmatrix} c_{\mathbf{k},\frac{1}{2}} \\ c^+_{-\mathbf{k},-\frac{1}{2}} \end{pmatrix} \,,$$

where the coefficients must satisfy the relation

$$|u_k|^2 + |v_k|^2 = 1 \,.$$

In terms of the new creation annihilation operators, we get

$$\mathcal{H}^0_{HFB} = \sum_{\mathbf{k}} \left\{ E_{\mathbf{k}} \left(c^+_{\mathbf{k},\frac{1}{2}} c_{\mathbf{k},\frac{1}{2}} + c^+_{\mathbf{k},-\frac{1}{2}} c_{\mathbf{k},-\frac{1}{2}} \right) + \mathscr{C}(\mathbf{k}) \right\} \,,$$

where

$$E(\mathbf{k}) = \sqrt{\varepsilon(\mathbf{k})^2 + |\Delta(\mathbf{k})|^2} \,,$$

and (in the infinite volume limit)

$$\Delta(\mathbf{k}) = -\frac{1}{(2\pi)^3} \int d\mathbf{k} \tilde{V}(\mathbf{k} - \mathbf{k}') \frac{\Delta(\mathbf{k}')}{2E(\mathbf{k}')} \left[1 - 2f\left(E(\mathbf{k}') \right) \right]$$

with the Fermi function

$$f(E) = \frac{1}{e^{\beta E} + 1} \,.$$

The HFB constant is

$$\mathscr{C}(\mathbf{k}) = \varepsilon_{\mathbf{k}} - E_{\mathbf{k}} + \frac{|\Delta_{\mathbf{k}}|^2}{2E_{\mathbf{k}}} \left[1 - 2f\left(E(\mathbf{k}') \right) \right] \,.$$

The chemical potential μ has to be determined from the equation for the average density of electrons, that in the new variables looks as

$$\frac{1}{(2\pi)^3} \int d\mathbf{k} \left[1 - \frac{\varepsilon(\mathbf{k})}{E(\mathbf{k})} [1 - 2f(E(\mathbf{k}))] \right] = \langle n \rangle .$$

The grand-canonical potential is

$$\mathscr{F} = -\frac{1}{\Omega\beta} \ln \left\{ Tr \left[e^{-\beta\mathscr{H}^0_{HFB}} \right] \right\} = \frac{1}{(2\pi)^3} \int d\mathbf{k} \left\{ \frac{2}{\beta} \ln [1 - f(E(\mathbf{k})] + \mathscr{C}(\mathbf{k}) \right\}$$

and the solution of the gap equation ensures also the extremum condition

$$\frac{\delta\mathscr{F}}{\delta\Delta(\mathbf{k})} = 0 ,$$

while the stability of the phase transition depends on the sign of its second derivative.

The gap equation obviously has a vanishing solution $\Delta(\mathbf{k}) = 0$ and possibly another one with $\Delta(\mathbf{k}) \neq 0$. If the solution with $\Delta(\mathbf{k}) \neq 0$ gives rise to a smaller grand-canonical potential, the anomalous solution is preferred, and we have a phase transition. In this case the new ground state is the new vacuum, with energy $\mathscr{C}(\mathbf{k})$, while the next excited state with a quasi-particle of wave-vector $\mathbf{k}$ has the energy $\mathscr{C}(\mathbf{k}) + E(\mathbf{k})$. The minimal distance between the two states may be finite and we have an energy gap for excitations. (Due to the rotational invariance, not only the unperturbed energies ε but also the new ones E, as well as Δ are functions only of $k \equiv |\mathbf{k}|$ and we may omit the vector notation.)

We have avoided to define the potential $V(\mathbf{x})$ itself and it is not at all clear if the BCS theory may be formulated explicitly as an attraction through a local potential in real space. Therefore, the above discussion serves only as a sketch of a possible scenario for a phase transition with a BCS-type electron correlation. For any given potential one may check (numerically) the expected properties.

Whether the above described theory or any similar one describes indeed a super-conductor may be checked only by understanding its electromagnetic properties. A theory of the resistance-less flow of currents is beyond any hope. Indeed, transport theory introduces elements of irreversibility (mostly by hand!) in order to get a finite conductivity. If one omits this step, then any model allows free flow of electric current. Then remains the question, what would impede dissipation? A vague idea about is the hope, that the existence of a gap above the Fermi energy might offer stability against weak perturbations. The only well defined possibility remains to look at least at magnetic properties in equilibrium.

7.4.1 Meissner Effect Within Equilibrium Linear Response

The simplest test of superconducting features in the above described model is to see whether it gives rise to the Meissner effect i.e. to expulsion of magnetic fields from the bulk? Since the Meissner effect occurs in equilibrium, one my use the equilibrium linear response theory (see Sect. 6.2).

The basic hamiltonian used in solid state theory omits terms of order $\frac{1}{c^2}$ and this means not only omitting relativistic corrections, but also ignoring velocity-velocity couplings between the electrons dictated by electrodynamics. Thus, the magnetic field produced by the electrons is completely absent. Therefore, we are compelled to interpret here the magnetic field as the "true" one being the sum of the external and the average internal magnetic fields in analogy to the treatment of longitudinal electric fields in a system of electrons without explicit Coulomb interaction. Nevertheless it was possible to compute the average s.c. longitudinal field produced by the electrons. An analogous possibility for the average electron current also is available.

We treat an infinite system and therefore the source of the magnetic field has to be a fictitious external current placed in the bulk. Let us consider a weak magnetic field that may be considered as a small perturbation. The peculiarity of the self-consistent approach here is, that the vector potential $\mathbf{A}(\mathbf{x})$ contained explicitly in the kinetic energy is not the only perturbation due to the magnetic field. The anomalous averages in the s.c. "Hamiltonian" $\mathscr{H}_{HFB}$ are also modified by the magnetic field. Therefore, the perturbation to consider for the linear response is

$$\mathscr{H}' = -\frac{1}{c}\int d\mathbf{x}\mathbf{j}(\mathbf{x})\mathbf{A}(\mathbf{x}) + \int d\mathbf{x}\int d\mathbf{x}'\, V(\mathbf{x}-\mathbf{x}')\left[\eta(\mathbf{x},\mathbf{x}')\psi_{-\frac{1}{2}}(\mathbf{x}')\psi_{\frac{1}{2}}(\mathbf{x}) + h.c.\right],$$

where

$$\eta(\mathbf{x},\mathbf{x}') \equiv \langle\psi_{\frac{1}{2}}^{+}(\mathbf{x})\psi_{-\frac{1}{2}}^{+}(\mathbf{x}')\rangle - \langle\psi_{\frac{1}{2}}^{+}(\mathbf{x})\psi_{-\frac{1}{2}}^{+}(\mathbf{x}')\rangle_0$$

is (the first order in the magnetic field) deviation of the averages from their unperturbed values. These must be simultaneously calculated self-consistently. We still must take into account also, that in then presence of a magnetic field the current density contains also a diamagnetic term

$$\mathbf{j}(\mathbf{x}) \equiv \frac{e}{2m}\psi^{+}(\mathbf{x})\left(-\imath\hbar\nabla + \frac{e}{c}\mathbf{A}(\mathbf{x})\right)\psi(\mathbf{x}) + h.c. \ .$$

One may use the equilibrium linear response formula of Sect. 6.2 with the previous results for the chosen model of attracting electrons to get the linear relationship between the average current and the vector potential.

From the current conservation it follows, that

$$\langle\tilde{j}_{\mu}(\mathbf{k})\rangle = \left(\delta_{\mu\nu} - \frac{k_{\mu}k_{\nu}}{k}\right)\kappa(k)\tilde{A}_{\nu}(\mathbf{k}) \ .$$

Due to the transversality of the vector potential in the Coulomb gauge this is equivalent to

$$\langle \tilde{j}_\mu(\mathbf{k}) \rangle = \kappa(k) \tilde{A}_\mu(\mathbf{k}).$$

The coefficient $\kappa(k)$ given by the equilibrium linear response theory has a rather
complicated expression in terms of the previously defined parameters and we omit
it. Actually, we are interested in this relationship only for small wave vectors (slowly
varying behavior in the coordinate space!). In the anomalous stable superconducting
phase having a non-vanishing gap $\Delta(\mathbf{k})$ at $\mathbf{k} = 0$ we get

$$\kappa(0) = -\frac{1}{c\Lambda}$$

with

$$\frac{1}{c\Lambda} = \frac{e^2}{m} \left[\langle n \rangle + \frac{4}{3} \frac{1}{(2\pi)^3} \int d\mathbf{p} \, (\varepsilon(\mathbf{p}) + \mu) \frac{\partial f(E(\mathbf{p}))}{\partial E(\mathbf{p})} \right].$$

It may be shown, that under the condition of the stability of the superconducting phase
Λ is indeed positive and the contribution produced by consideration of $\eta(\mathbf{x}, \mathbf{x}')$ to this
result vanishes. Therefore, the above described model seems to produce a Meissner
effect, since in the Coulomb gauge the second London equation reads as

$$\mathbf{j} = -\frac{1}{c\Lambda}\mathbf{A} .$$

However, a word of caution is compulsory. The prospect of a correct understanding
of the Meissner effect in the frame of any hamiltonian theory of electrons is still
not convincing. Such a theory, as we already mentioned, ignores the magnetic field
produced by the electrons themselves. This approximation is not dangerous whenever
this magnetic field is negligible as compared to the external field. However, this is
not the case by the Meissner effect. Since in equilibrium, with a stationary average
current the retardation dictated by electromagnetic theory may be omitted, we have
an internal vector potential produced by the average internal transverse current

$$\mathbf{A}^{int}(\mathbf{x}) = \int d\mathbf{x}' \frac{\langle \mathbf{j}_\perp^{int}(\mathbf{x}') \rangle}{c|\mathbf{x} - \mathbf{x}'|} .$$

However we cannot compare this field to the external one (absent in this formulation)
in order to check that indeed the diamagnetic currents screen the external magnetic
field. Here we might be confronted with the limits of today's solid state theory.

Chapter 8
Low Dimensional Semiconductors

Quasi two-dimensional semiconductor layers offer a lot of spectacular prop-
erties of which the most famous is the Quantum Hall Effect. We limit
ourselves, however only to the presentation of the simplest, well under-
stood, but nevertheless surprising peculiarities of the classical and quantum-
mechanical two-dimensional motion of Coulomb interacting electrons in the
presence of a strong transverse magnetic field.

In the last decades progresses in semiconductor technology produced ultra-thin semiconductor layer systems in which by a suitable choice of the layers, the electrons and holes are restricted to a two dimensional motion. An ultra-thin layer of a semiconductor is inserted between two thick layers of another semiconductor with a larger band-gap and therefore, both the electrons and holes in the ultra-thin layer are in a quantum well and their lowest states are discrete as it is shown in Fig. 8.1.

At sufficiently low temperatures the electrons and holes sit on their lowest levels (shown here in red and green). It means, the transverse motion at low temperatures is "frozen" in its lowest lying state $\phi_0(z)$ in the potential well created by the adjoining layers and the wave function of an electron, in a good approximation is given by

$$\Psi(x, y, z) = \psi(x, y)\phi_0(z) ,$$

with the coordinates x, y lying in the plane and z being the transverse coordinate. This is the experimental realization of a two dimensional (2D) electron-hole system, which has a lot of very interesting properties.

Already the free motion in 2D has a peculiarity, namely the one-electron state density is constant above $\varepsilon = 0$

$$z(\varepsilon) = \frac{1}{(2\pi)^2} \int d\mathbf{k}\delta(\varepsilon - \frac{\hbar^2 k^2}{2m}) = \frac{m}{2\pi\hbar^2}\theta(\varepsilon) .$$

© Springer International Publishing AG, part of Springer Nature 2018
L. A. Bányai, *A Compendium of Solid State Theory*,
https://doi.org/10.1007/978-3-319-78613-1_8

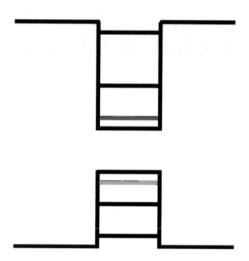

Such systems show above all spectacular properties in the presence of a strong
magnetic field transverse to the plane. The most famous one is the Quantum Hall
Effect. We will not try to describe here the different, sometimes contradicting theories
of this effect. Nevertheless, we want to bring the attention to the kind of strange
physics we encounter in two dimensions (2D).

8.1   Exciton in 2D

Let us consider an electron-hole pair with Coulomb attraction in a 2D semiconductor.
In the effective mass approximation, the relative motion in cylindrical coordinates is
described by the Hamiltonian

$$H = -\frac{\hbar^2}{2m}\frac{1}{\rho}\frac{\partial}{\partial\rho}\left(\rho\frac{\partial}{\partial\rho}\right) - \frac{1}{\rho^2}\frac{\partial^2}{\partial\phi^2} + \frac{e^2}{\varepsilon\rho},$$

where m is the reduced mass of the pair and ε is the dielectric constant (supposed
here as being the same in the 2D layer as in the surrounding semiconductor). The
eigenstates of this Hamiltonian are characterized by two quantum numbers $n =
0, 1, 2\ldots$ and $\mu = \pm0, \pm1, \pm2, \ldots$. The lowest eigenstate ($n = 0$, $\mu = 0$) is

$$\psi_{0,0} = \frac{4}{\sqrt{2\pi}\,a}\,e^{-\frac{2\rho}{a_B}},$$

while the ground state energy is

$$E_{0,0} = -4E_R.$$

where we used as parameter the Bohr radius a_B and the Rydberg energy E_R of the 3D exciton. A comparison with the ground state exciton wave function in 3D (see Sect. 3.1.1) shows that the exciton radius in 2D is twice larger than in 3D and its binding energy is four times bigger. Contrary to any expectations, the transversely compressed exciton shrinks also in the still allowed two dimensions.

8.2 Motion of a 2D Electron in a Strong Magnetic Field

According to the discussed 3D motion of an electron in a homogeneous magnetic field (see Sect. 2.2.2) the stationary states in the Landau gauge are given by a plane-wave in the field direction (along the z axis) and in the transverse plane by the Landau states of discrete energies. If one restricts the motion to the plane x,y, then the energy of such a Landau state is just

$$\varepsilon_{n,X,} = \hbar\omega_c(n + \frac{1}{2}) \qquad (n = 0, 1, 2, ...) .$$

Since these energies do not depend on the quantum number X (the x-coordinate of the center of the cyclotron motion), they are (in the absence of boundary conditions!) infinitely degenerate.

In what follows we consider very strong magnetic fields at very low temperatures and therefore we may consider, that all the spins are aligned along the magnetic field. We shall consider the motion of such a 2D electron in a strong magnetic field perpendicular to the plane of motion in the presence of an external potential $U(\mathbf{r})$.

The simplest approach to consider is to ignore the higher lying Landau levels and, if the potential is weak, to consider its projection on the lowest lying Landau state $(n = 0)$ within first order perturbation for the energy

$$E(X) = \frac{1}{2}\hbar\omega_0 + \langle 0, X|U(x, y)|0, X\rangle .$$

Kubo's more profound approach considers the limit of ultra-strong magnetic fields for arbitrary potentials and an arbitrary kinetic energy T. The 2D Hamilton operator is

$$H = T(\boldsymbol{\pi}) + U(\mathbf{r}) ,$$

where the generalized momenta $\boldsymbol{\pi}$ in the presence of the vector potential $\mathbf{A}$ are

$$\boldsymbol{\pi} \equiv \mathbf{p} + \frac{e}{c}\mathbf{A}(\mathbf{r}) ; \qquad \mathbf{p} \equiv \imath\hbar\nabla .$$

Within the set of operators $(\boldsymbol{\pi}, \mathbf{r})$ the only non-vanishing commutators are

$$[\pi_x, \pi_y] = \frac{\hbar e}{\imath c} B; \quad [\pi_x, x] = [\pi_y, y] = \frac{\hbar}{\imath}$$

Defining new operators

$$\xi = \frac{c}{eB}\pi_y, \quad \eta = \frac{c}{eB}\pi_x$$

and

$$X = x - \xi, \quad Y = y - \eta$$

it results, that

$$[\xi, \eta] = \frac{\hbar c}{\imath e B} \equiv \frac{l_B^2}{\imath},$$

$$[X, Y] = -\frac{\hbar c}{\imath e B} \equiv -\frac{l_B^2}{\imath},$$

while the other commutators of the new operators vanish. The kinetic energy T depends only on the new operators ξ and η. Also, in the absence of the potential U the operators X and Y do not change in time. They are assimilated with the coordinates of the cyclotron motion, while the operators ξ and η are the relative coordinates.

In the presence of the external potential, either in the frame of Quantum Mechanics or within classical mechanics, the center of the cyclotron motion moves according to

$$\dot{X} = \frac{c}{eB}\frac{\partial U}{\partial y}, \quad \dot{Y} = -\frac{c}{eB}\frac{\partial U}{\partial x}$$

According to the above canonical commutation rules one has the uncertainty relations

$$\triangle X \triangle Y = 2\pi l_B^2$$

and since ultra-strong magnetic fields imply $l_B \to 0$, a classical description with classical coordinates X, Y is appropriate. On the other hand, if the potential is slowly varying on the scale of the magnetic length, one may ignore the relative motion and to a very good approximation one could use the classical equations of motion

$$\dot{X} = \frac{c}{eB}\frac{\partial U(X, Y)}{\partial Y}, \quad \dot{Y} = -\frac{c}{eB}\frac{\partial U(X, Y)}{\partial X}.$$

It follows, that

$$\frac{\partial Y}{\partial X} = \frac{\dot{Y}}{\dot{X}} = -\frac{\frac{\partial U(X,Y)}{\partial X}}{\frac{\partial U(X,Y)}{\partial Y}}.$$

However, this corresponds to the motion on a curve defined by $U(x, y) = U_0$. Indeed, by using the definition of the implicit derivative one gets

$$\frac{\partial y}{\partial x} = -\frac{\frac{\partial U(x,y)}{\partial x}}{\frac{\partial U(x,y)}{\partial y}}.$$

To conclude, in the limit of ultra-strong magnetic fields the motion of the cyclotron center of electrons in 2D with arbitrary kinetic energy in the presence of an external potential is just a classical one along the equipotential curves of the potential.

8.3 Coulomb Interaction in 2D in a Strong Magnetic Field

8.3.1 Classical Motion

A stranger aspect of the 2D motion in a strong magnetic field is how Coulomb forces act. Let us consider first the classical problem of the motion of two particles of opposite charges (electron and hole). We are interested only in the relative motion, therefore one of the particles we may keep fixed in the origin. A numerical solution of the corresponding Newton equations shows the trajectory in Fig. 8.2. This is a somewhat complicated picture, but the particles, as expected seem to attract each other and stick together.

To big surprise, even two identically charged particles in 2D, having repulsive Coulomb forces, stick together, showing an effective attraction, as it is illustrated on Fig. 8.3. Of course, the Coulomb force tries to accelerate the electrons in the repulsive manner, but the accelerated electron is returned by the bending in the magnetic field. The only escape would have been in the now forbidden transverse direction.

If one considers a whole cluster of electrons these are sticking together in a cluster, as it is shown in Fig. 8.4.

8.3.2 Quantum Mechanical States

The quantum-mechanical analysis of the motion of two Coulomb repulsive particles confirms also the existence of bound electron-electron states in 2D.

The quantum-mechanical Hamiltonian for the 3D relative motion of two electrons in the presence of a magnetic field $\mathbf{B}$ is

$$H^{3D} = -\frac{\hbar^2}{2m}\nabla^2 - \frac{e\hbar}{mc}\imath\left(\mathbf{A}(\mathbf{r}) \cdot \nabla + \frac{1}{2}\nabla\mathbf{A}(\mathbf{r})\right) + \frac{e^2}{2mc^2}\mathbf{A}(\mathbf{r})^2 + \frac{e^2}{r}.$$

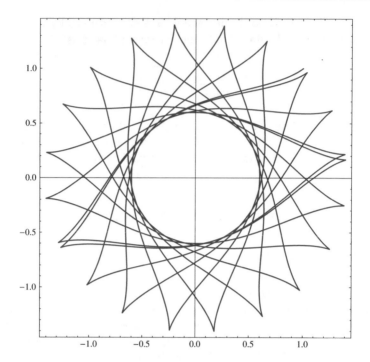

Fig. 8.2 Classical relative motion of opposite charged particles in 2D in the presence of a transverse magnetic field

If one chooses the divergence-less vector potential (with the magnetic field in the z-direction)

$$\mathbf{A}(\mathbf{r}) \equiv \left(-\frac{1}{2}By, \, \frac{1}{2}Bx, \, 0 \right)$$

one may write the 2D Hamiltonian describing the in-plane motion in cylinder coordinates ρ, ϕ as

$$H^{2D} = -\frac{\hbar^2}{2m} \frac{1}{\rho} \frac{\partial}{\partial \rho} \left(\rho \frac{\partial}{\partial \rho} \right) - \frac{1}{\rho^2} \frac{\partial^2}{\partial \phi^2} - \iota \frac{e\hbar B}{2mc} \frac{\partial}{\partial \phi} + \frac{e^2 B^2}{8mc^2} \rho^2 + \frac{e^2}{\rho} \ .$$

One looks, as usual, for the eigenfunctions as

$$\psi(\rho, \phi) = u(\rho) e^{\iota \mu \phi} \qquad (\mu = 0, 1, 2, \ldots) \ .$$

Then the eigenvalue problem for the radial part is:

$$-\frac{\hbar^2}{2m} \frac{1}{\rho} \frac{\partial}{\partial \rho} \left(\rho \frac{\partial u}{\partial \rho} \right) + \left(\frac{\hbar^2}{2m} \frac{\mu^2}{\rho^2} + \frac{e\hbar B}{2mc} \mu + \frac{e^2 B^2}{8mc^2} \rho^2 + \frac{e^2}{\rho} \right) u = E_\mu u \ .$$

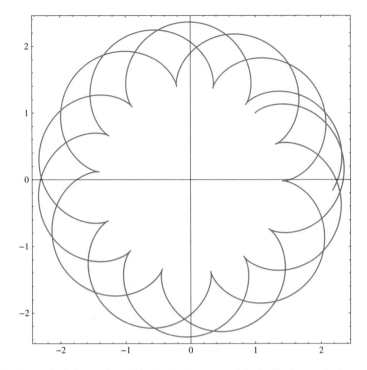

Fig. 8.3 Classical relative motion of identically charged particles in 2D electron in the presence of a transverse magnetic field

This Schroedinger equation differs (up to a shift in the energy with $\frac{e\hbar B}{2mc}\mu$) from that of a radial 2D oscillator just due to the Coulomb term $\frac{e^2}{\rho}$. Since in a very strong magnetic field the spins are supposed to be aligned along the magnetic field and the wave functions must be anti-symmetrical for fermions, only odd angular momenta μ are of interest. In terms of the dimensionless parameter

$$\xi = \frac{eB}{2c\hbar}\left(\frac{\hbar^2}{2me^2}\right)^2$$

an ultra-strong magnetic field corresponds to $\xi \gg 1$. The radial wave function of the lowest $\mu = 1$ state of the 2D oscillator is

$$R_{1,0}(\rho) = \frac{\xi}{\sqrt{\pi}}\rho e^{-\frac{1}{2}\xi\rho^2}; \qquad 2\pi\int_0^\infty \rho|R(\rho)|^2 d\rho = 1 \ ,$$

where for convenience the radius ρ is measured in units of the length $\lambda = \frac{\hbar^2}{2me^2}$.

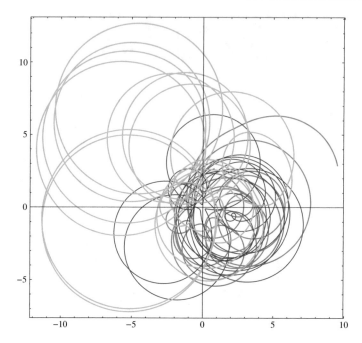

Fig. 8.4 Classical motion of 4 electrons (one is kept fixed in the origin) in 2D in the presence of the magnetic field

A numerical solution of the lowest eigenfunction including Coulomb repulsion for $\mu = 1$ and $\xi = 10$ is shown red in Fig. 8.5, while the corresponding oscillator function is shown in blue.

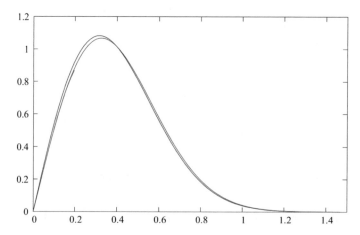

Fig. 8.5 The lowest $\mu = 1$ eigenfunction with and without Coulomb repulsion

The two wave functions are surprisingly close to each other. The contribution of the Coulomb potential to the energy may be approximated by first order perturbation theory as its average over the oscillator function. The result again lies surprisingly close to the exact value (in our example, up to four digits).

On the other hand, without the Coulomb potential the wave function would have been centered at any arbitrary position in the plane. The existence of the repulsive center however fixes its position ! This suggested the construction of the so called Laughlin-wave functions for many electrons out of oscillator wave functions.

It is worth to remark also, that the lowest s-wave ($\mu = 0$) state shows in concordance with the formerly discussed theory of Kubo a motion concentrated on a circle (equipotential line for the Coulomb potential).

Chapter 9
Shortcut of Theoretical Physics

Here we offer a list of definitions and formulas, that may help the reader to refresh his knowledge of theoretical physics.

9.1 Classical Mechanics

Lagrange function of a point-like particle in the presence of a potential $U(\mathbf{x}, t)$:

$$L(\mathbf{x}, \dot{\mathbf{x}}) = \frac{m}{2}\dot{\mathbf{x}}^2 - U(\mathbf{x}, t))$$

Euler equations:

$$\frac{\partial}{\partial t}\frac{\partial L}{\partial \dot{\mathbf{x}}} - \frac{\partial L}{\partial \mathbf{x}} = 0$$

Hamilton function:

$$\mathbf{q} = \mathbf{x}; \quad \mathbf{p} = \frac{\partial L}{\partial \dot{\mathbf{x}}}$$

$$H(\mathbf{p}, \mathbf{q}) = -L + \mathbf{p}\dot{\mathbf{q}}$$

Poisson brackets:

$$\{f, g\} = \frac{\partial f}{\partial \mathbf{q}}\frac{\partial g}{\partial \mathbf{p}} - \frac{\partial g}{\partial \mathbf{q}}\frac{\partial f}{\partial \mathbf{p}}$$

© Springer International Publishing AG, part of Springer Nature 2018
L. A. Bányai, *A Compendium of Solid State Theory*,
https://doi.org/10.1007/978-3-319-78613-1_9

Equations of motion:

$$\dot{q} = \{H, q\}; \quad \dot{p} = \{H, p\}$$

The Hamiltonian formulation may be extended to include the interaction with a magnetic field, leading to the velocity dependent Lorentz force! In this case the velocity is related to the canonical momentum by

$$m\dot{x} = p + \frac{e}{c}A(x, t) \ .$$

9.2   One-Particle Quantum Mechanics

State of the particle (wave function):

$$\psi(x, t)$$

Time evolution by the Schroedinger equation:

$$i\hbar\frac{\partial}{\partial t}\psi = H(t)\psi$$

Hamilton operator in a classical electromagnetic field described by the scalar and vector potentials $V(x, t), A(x.t)$:

$$H(t) = \frac{1}{2m}\left(-i\hbar\nabla + \frac{e}{c}A(x, t)\right)^2 + eV(x, t)$$

Scalar product:

$$(\psi_1, \psi_2) \equiv \int dx\,\psi_1(x)^*\psi_2(x)$$

Matrix elements of an operator $\mathscr{A}$ are defined by

$$(\psi_1, \mathscr{A}\psi_2) \equiv \int dx\,\psi_1(x)^*\mathscr{A}\psi_2(x) = \int dx\,(\mathscr{A}^+\psi_1(x))^*\psi_2(x)$$

Here $\mathscr{A}^+$ is the adjoint of $\mathscr{A}$. Observables are self-adjoint:

$$\mathscr{A} = \mathscr{A}^+$$

Averages:

$$\langle\mathscr{A}\rangle \equiv \int dx\,\psi(x)^*\mathscr{A}\psi(x)$$

Stationary problem in the presence of a potential:

$$H = -\frac{\hbar^2}{2m}\nabla^2 + eU(\mathbf{x})$$

Eigenvalue problem:

$$H\phi_i = E_i\phi_i$$

Orthonormalization and completeness:

$$(\phi_i, \phi_j) = \delta_{ij}$$
$$\sum_i \phi_i(x)\phi_i(x')^* = \delta(x, x')$$

9.2.1 Dirac's "bra/ket" Formalism

States are represented by

bra:
$$< \psi|$$

ket:
$$|\psi >$$

Scalar product (*bracket*):
$$< \psi_1|\psi_2 >$$

Eigenstates and eigenvalues:
$$H|i >= E_i|i >$$

Projectors on an eigenstate:
$$|i >< i|$$

$$\sum_i |i >< i| = 1$$

$$H = \sum_i E_i|i >< i|$$

9.3   Perturbation Theory

9.3.1   Stationary Perturbation

Perturbation proportional to a small λ:

$$H = H_0 + \lambda H'$$

Expansion in λ:

$$H_0 \phi_n^{(0)} = E_n^{(0)} \phi_n^{(0)}; \; H\phi_n = E_n\phi_n$$
$$\phi_n = \phi_n^{(0)} + \lambda \phi_n^{(1)} + \lambda^2 \phi_n^{(2)} \ldots$$
$$E_n = E_n^{(0)} + \lambda E_n^{(1)} + \lambda^2 E_n^{(2)} + \ldots$$

$$H'_{mn} \equiv (\phi_m^{(0)}, H'\phi_n^{(0)})$$

Without degeneracy:

$$\phi_n^{(1)} = \sum_{m(\neq n)} \frac{H'_{mn}}{E_n^{(0)} - E_m^{(0)}}$$

$$E_n^{(1)} = H'_{nn}$$

$$E_n^{(2)} = \sum_{m(\neq n)} \frac{|H'_{mn}|^2}{E_n^{(0)} - E_m^{(0)}}$$

With degeneracy:

$$E_{n,s}^{(0)} = E_n^{(0)}; \; (s = 1, \ldots S)$$

to zeroth order for the eigenstate and first order in the eigenenergy

$$\sum_{s'=1}^{S} H'_{ns,ns'} \phi_{n,s'}^{(0)} = E^{(1)} \phi_{n,s}^{(0)}$$

9.3.2   Time Dependent Adiabatic Perturbation

$$H(t) = H_0 + \lambda H'(t)$$

Schroedinger equation

$$i\hbar \frac{\partial}{\partial t} |\psi(t)> = H(t)|\psi(t)>$$

Interaction picture

$$|\psi(t)>_I \equiv e^{\frac{i}{\hbar}H_0 t}|\psi(t)>$$

$$i\hbar\frac{\partial}{\partial t}|\psi(t)>_I = \lambda e^{\frac{i}{\hbar}H_0 t}H'(t)e^{-\frac{i}{\hbar}H_0 t}|\psi(t)>_I = \lambda H'(t)_I|\psi(t)>_I$$

Unitary evolution in the interaction picture

$$U(t,t_0) = T\left\{e^{-\frac{i}{\hbar}\lambda\int_{t_0}^{t}H'(t')_I dt'}\right\}$$

$T\{\ldots\}$ *indicates that the operators* $H'(t)_I$ *in the expansion of the exponential should be ordered according to chronology.*

S-Matrix

$$S = \lim_{t\to\infty}\lim_{t_0\to-\infty}U(t,t_0)$$

Asymptotic transition rate between the unperturbed eigenstates $|n>$ and $|m>$ of H_0

$$\mathscr{W}_{nm} \equiv \lim_{t\to\infty}\lim_{t_0\to-\infty}\frac{d}{dt}|<n|U(t,t_0)|m>|^2$$

The "golden rule" to second order in λ, with an adiabatic, but oscillating in time perturbation (light absorption)

$$H'(t) = (H'e^{i\omega t} + h.c.)e^{-0|t|}$$

$$\mathscr{W}_{nm} = \frac{2\pi}{\hbar}\lambda^2|H'_{nm}|^2\delta(E_n^{(0)} - E_m^{(0)} \pm \hbar\omega)$$

9.4 Many-Body Quantum Mechanics

9.4.1 Configuration Space

Hamilton operator and wave function of Coulomb interacting particles in configuration space:

$$H^{(N)} = \sum_{i=1}^{N}\left(-\frac{\hbar^2}{2m}\nabla_i^2 + U(\mathbf{x}_i)\right) + \frac{1}{2}\sum_{i\neq i'}^{N}\frac{e^2}{|\mathbf{x}_i - \mathbf{x}_{i'}|}$$

$$H^{(N)}\Phi(x_1,\cdots x_N) = E_N\Phi(x_1,\cdots x_N)$$

$\Phi(x_1,\cdots x_N)$ *antysymmetrical for fermions and symmetrical for bosons!*

9.4.2   Fock Space (Second Quantization)

9.4.2.1   Fermions

Vacuum state:
$$|0 >$$

Creation and annihilation operators:
$$c^+|0 >= |1 > \qquad c^+|1 >= 0$$
$$c|1 >= |0 > \qquad c|0 >= 0$$

Occupation number operator:
$$c^+c|0 >= 0; \quad c^+c|1 >= |1 >$$

To every one-particle state $|k >$ corresponds a creation operator: c_k^+.

Anti-commutation:
$$\left[c_{k'}, c_k^+\right]_+ = \delta_{k,k'}$$

Many fermion state:
$$|\Phi >= c_1^+ c_2^+ \cdots c_N^+ |0 >$$

$$c_k^+|\cdots n_k \cdots > = (-1)^{\nu_k}\sqrt{1-n_k} \ |\cdots n_k+1 \cdots >,$$
$$c_k|\cdots n_k \cdots > = (-1)^{\nu_k}\sqrt{n_k} \ |\cdots n_k-1 \cdots >$$

$$\nu_k = \sum_{i<k} n_i$$

Second quantized wave function:
$$\psi(x) = \sum_k \phi_k(x)c_k$$
$$\left[\psi(x), \psi(x')^+\right]_+ = \delta(x-x')$$
$$\left[\psi(x), \psi(x')\right]_+ = 0$$

Vacuum state (*ground state*):
$$|0 >$$

Basis:

$$|x_N, \cdots, x_1 > \equiv \psi(x_1)^+ \cdots \psi(x_N)^+ |0 >$$

$$< x_1, \cdots, x_N | x_1', \cdots, x_N' > = det \left| \delta(x_i - x_j') \right|$$

Hamilton operator and state of Coulomb interacting fermions in Fock space:

$$H = \int dx \psi(x)^+ \left[-\frac{\hbar^2}{2m} \nabla^2 + U(x) \right] \psi(x) + \frac{1}{2} \int dx \int dx' \psi(x)^+ \psi(x')^+ \frac{e^2}{|x - x'|} \psi(x') \psi(x)$$

The relationship to the configuration space description:

$$H |\Phi_N > = E_N |\Phi_N >$$

$$|\Phi_N > \equiv \int dx_1 \cdots \int dx_N \frac{1}{\sqrt{N!}} \Phi(x_1, \cdots x_N) |x_1, \cdots, x_N >$$

Particle density:

$$n(\mathbf{x}) = \psi(\mathbf{x})^+ \psi(\mathbf{x}$$

Current density:

$$\mathbf{j}(\mathbf{x}) = -\frac{\iota \hbar}{2m} \psi(\mathbf{x})^+ \nabla \psi(\mathbf{x}) + h.c.$$

Continuity equation.

$$\frac{\partial n(\mathbf{x})}{\partial t} + \nabla \mathbf{j}(\mathbf{x}) = 0$$

9.4.2.2 Bosons

Creation and annihilation operators:

$$a|0 > = 0$$

$$a^+ |n > = \sqrt{n+1} |n+1 >$$

$$a|n > = \sqrt{n} |n-1 >$$

$$a^+ |n > = \sqrt{n+1} |n+1 >$$

$$a|n > = \sqrt{n} |n-1 >$$

Occupation number operator:

$$a^+ a |n > = n |n >$$

Commutation:

$$\left[a_{k'}, a_k^+\right] = \delta_{k,k'}$$

Many boson state:

$$\Phi = \frac{1}{\sqrt{n_1! \ldots n_N!}} (a_1^+)^{n_1} \ldots (a_N^+)^{n_N} |> \qquad \sum_i n_i = N$$

Second quantized wave function:

$$\psi(x) = \sum_k \phi_k(x) a_k$$

$$\left[\psi(x), \psi(x')^+\right] = \delta(x - x')$$
$$\left[\psi(x), \psi(x')\right] = 0$$

$$|x_N, \cdots, x_1 > \equiv \psi(x_1)^+ \cdots \psi(x_N)^+ |0>$$

$$< x_1, \ldots, x_N | x_1', \ldots, x_N' > = \sum_P \delta(x_1 - x_{j_1}') \ldots \delta(x_N - x_{j_N}')$$

9.5   Density Matrix (Statistical Operator)

Pure quantum mechanical ensemble defined by the state Φ.

$$\langle \mathscr{A} \rangle = < \Phi | \mathscr{A} | \Phi >$$

Mixed ensemble:

$$|\Phi_\alpha > \qquad \text{with   probability }\; p_\alpha \qquad (0 \le p_\alpha \le 1; \;\; \sum_\alpha p_\alpha = 1)$$

$$\langle \mathscr{A} \rangle = \sum_\alpha p_\alpha < \Phi_\alpha | \mathscr{A} | \Phi_\alpha >$$

Density matrix (Statistical operator):

$$R \equiv \sum_\alpha p_\alpha |\Phi_\alpha > < \Phi_\alpha|$$

$$Tr\{R\} = 1 ; \quad R > 0$$

$$\iota\hbar\frac{\partial R}{\partial t} = [H, R]$$

$$\langle\mathscr{A}\rangle = Tr\{\mathscr{A}R\}$$

Micro-canonical equilibrium:

$$R_0 = \delta(H^{(N)} - E_N)$$

Grand-canonical equilibrium.

$$R_0 = \frac{e^{-\beta(H-\mu N)}}{Tr\left(e^{-\beta(H-\mu N)}\right)}$$

Grand-canonical potential:

$$\mathscr{F}(V, T, \mu) == -k_B T \ln Z ; \qquad Z \equiv Tr\left\{e^{-\beta(H-\mu N)}\right\}$$

free fermions:

$$\langle a_k^+ a_k\rangle = \frac{1}{e^{\beta(e_k-\mu)} + 1}$$

free bosons:

$$\langle a_k^+ a_k\rangle = \frac{1}{e^{\beta(e_k-\mu)} - 1}$$

9.6   Charged Particles and Electromagnetic Fields

Classical Maxwell-Lorenz equations

$$\nabla \times \mathbf{b} = \frac{4\pi}{c}\mathbf{j} + \frac{1}{c}\frac{\partial}{\partial t}\mathbf{e}$$

$$\nabla \times \mathbf{e} = -\frac{1}{c}\frac{\partial}{\partial t}\mathbf{b}$$

$$\nabla\mathbf{b} = 0$$

$$\nabla\mathbf{e} = 4\pi\rho$$

Continuity equation:

$$\nabla\mathbf{j} + \frac{\partial}{\partial t}\rho = 0$$

Sources (*including external ones without dynamics*):

$$\rho(\mathbf{r}, t) = \sum_i e_i \delta(\mathbf{r} - \mathbf{r}_i(t)) + \rho^{ext}(\mathbf{r}, t)$$

$$\mathbf{j}(\mathbf{r}, t) = \sum_i e_i \mathbf{v}_i(t) \delta(\mathbf{r} - \mathbf{r}_i(t)) + \mathbf{j}^{ext}(\mathbf{r}, t)$$

$$\nabla \mathbf{j}^{ext} + \frac{\partial \rho^{ext}}{\partial t} = 0$$

Electromagnetic potentials:

$$\mathbf{b} = \nabla \times \mathbf{a}$$

$$\mathbf{e} = -\nabla v - \frac{1}{c} \frac{\partial}{\partial t} \mathbf{a}$$

Gauge invariance:

$$v \to v + \frac{1}{c} \frac{\partial}{\partial t} \Lambda$$

$$\mathbf{a} \to \mathbf{a} - \nabla \Lambda$$

Equations of the potentials:

$$-\nabla^2 \mathbf{a} + \nabla(\nabla \mathbf{a}) + \frac{1}{c^2} \frac{\partial^2}{\partial t^2} \mathbf{a} = \frac{4\pi}{c} \mathbf{j} - \frac{1}{c} \nabla \frac{\partial}{\partial t} v$$

$$-\nabla^2 v - \frac{1}{c^2} \nabla \frac{\partial}{\partial t} \mathbf{a} = 4\pi \rho$$

Coulomb gauge:

$$\nabla \mathbf{a} = 0$$

$$-\nabla^2 v = 4\pi \rho$$

$$-\nabla^2 \mathbf{a} + \frac{1}{c^2} \frac{\partial^2}{\partial t^2} \mathbf{a} = \frac{4\pi}{c} \mathbf{j} - \frac{1}{c} \nabla \frac{\partial}{\partial t} v \equiv \frac{4\pi}{c} \mathbf{j}_\perp$$

Transverse current:

$$\nabla \mathbf{j}_\perp = 0$$

$$\mathbf{j}_\perp(\mathbf{r}, t) \equiv \mathbf{j}(\mathbf{r}, t) + \frac{1}{4\pi} \nabla \int d\mathbf{r}' \frac{\nabla' \mathbf{j}(\mathbf{r}', t)}{|\mathbf{r} - \mathbf{r}'|}$$

Solution (with $\nabla \mathbf{a}^{ext} = 0$):

$$v(\mathbf{r}, t) = \int d\mathbf{r}' \frac{\rho^{int}(\mathbf{r}', t)}{|\mathbf{r} - \mathbf{r}'|} + V^{ext}(\mathbf{r}, t)$$

$$\mathbf{a}(\mathbf{r}, t) = \int d\mathbf{r}' \frac{\mathbf{j}^{int}_\perp (\mathbf{r}', t - \frac{1}{c}|\mathbf{r} - \mathbf{r}'|)}{c|\mathbf{r} - \mathbf{r}'|} + \mathbf{a}^{ext}(\mathbf{r}, t)$$

Newton's equation of motion with Lorentz forces:

$$m_i \frac{d}{dt} \mathbf{v}_i = e_i \left(\mathbf{e}'(\mathbf{r}_i, t) + \frac{1}{c} \mathbf{v}_i \times \mathbf{b}'(\mathbf{r}_i, t) \right)$$

Here $\mathbf{e}'$ and $\mathbf{b}'$ are the fields without self-action!
For $\frac{v}{c} \to 0$, one ignores the magnetic field created by the particles themselves $\mathbf{a}^{int}$ and the particle motion may be separated:

$$H = \sum_i \left(\frac{1}{2m_i} (\mathbf{p}_i + \frac{e_i}{c} \mathbf{a}^{ext}(\mathbf{r}_i, t))^2 + e_i V^{ext}(\mathbf{r}_i, t) \right) + \frac{1}{2} \sum_{i \neq j} \frac{e_i e_j}{|\mathbf{r}_i - \mathbf{r}_j|} .$$

In this standard approximation used in solid state theory, the magnetic field created by the charged particles is ignored!
Macroscopic fields as averages:

$$\mathbf{E} = \langle \mathbf{e} \rangle; \qquad \mathbf{B} = \langle \mathbf{b} \rangle$$

Macroscopic Maxwell equations.

$$\nabla \times \mathbf{B} = \frac{4\pi}{c}(\langle \mathbf{j} \rangle + \mathbf{j}^{ext}) + \frac{1}{c} \frac{\partial}{\partial t} \mathbf{E}$$

$$\nabla \times \mathbf{E} = -\frac{1}{c} \frac{\partial}{\partial t} \mathbf{B}$$

$$\nabla \mathbf{B} = 0$$

$$\nabla \mathbf{E} = 4\pi(\langle \rho \rangle + \rho^{ext})$$

The classical macroscopic theory of electromagnetism relates the average sources ($\langle \rho \rangle$, $\langle \mathbf{j} \rangle$) to the fields ($\mathbf{E}$, $\mathbf{B}$) by certain phenomenological relationships and leaves the foundation of these relationships to the microscopical theories.

The non-relativistic quantum-mechanical Hamiltonian of charged particles in external electric and magnetic fields:

$$H = \sum_{\sigma,\sigma'} \sum_{\alpha} \int d\mathbf{r} \Psi_{\alpha,\sigma}(\mathbf{r})^+ \left\{ \frac{1}{2m_\alpha} \left(-\imath\hbar\nabla + + \frac{e_\alpha}{c} \mathbf{A}^{ext}(\mathbf{r},t) \right)^2 \delta_{\sigma,\sigma'} \right.$$

$$\left. + e_\alpha V^{ext}(\mathbf{r},t)\delta_{\sigma,\sigma'} + \mu_B \boldsymbol{\sigma}_{\sigma\sigma'} \mathbf{B}(\mathbf{r},t)\Psi_{\alpha,\sigma'}(\mathbf{r}) \right\}$$

$$+ \frac{1}{2} \sum_{\sigma\sigma'} \sum_{\alpha,\alpha'} \int d\mathbf{r} \int d\mathbf{r}' \Psi_{\alpha,\sigma}(\mathbf{r})^+ \Psi_{\alpha',\sigma'}(\mathbf{r}')^+ \frac{e_\alpha e_{\alpha'}}{|\mathbf{r}-\mathbf{r}'|} \Psi_{\alpha',\sigma'}(\mathbf{r}')\Psi_{\alpha,\sigma}(\mathbf{r}) \quad .$$

Here the index α runs over the different sorts of charged particles, while σ runs over the spin projections (preferentially on the magnetic field direction).

The charge and current density operators:

$$\rho(\mathbf{x}) = \sum_{\sigma} \sum e_\alpha \Psi_{\alpha,\sigma}(\mathbf{r})^+ \Psi_{\alpha,\sigma}(\mathbf{r})$$

$$\mathbf{j}(\mathbf{x}) = \sum_{\sigma} \sum_{\alpha} \frac{e_\alpha}{2} \Psi_{\alpha,\sigma}(\mathbf{r})^+ \left(-\imath\hbar\nabla + \frac{e_\alpha}{c} \mathbf{A}^{ext}(\mathbf{r},t) \right) \Psi_{\alpha,\sigma}(\mathbf{r}) + h.c. \ .$$

The magnetic moment of the spins has to be included separately! It is sometimes useful to consider the quantum mechanical nature of the electromagnetic field (photons), while leaving the description of the charged particles in the former described quantum-mechanical, but non-relativistic description.

However, a consistent relativistic and quantum-mechanical formulation of the electrodynamics of particles and fields may be formulated only in the frame of the quantum electrodynamics. That is much beyond the frame of today's solid state theory.

Chapter 10
Home-Work

For all readers it is a useful exercise to perform the omitted details of the proofs. Them interested reader I recommend some lengthy but useful home-works an ambitious graduate student should nevertheless be able to perform successfully. I would like to encourage the reader also to write their own programs and play through different funny scenarios around the examples given in the Sects. 2.6.2, 5.2 and 8.3.1.

10.1 The Kubo Formula

Start from the linear response formula of Sect. 6.1 with the perturbation caused by coupling to (time dependent) electromagnetic potentials

$$H'(t) = \int d\mathbf{x} \left\{ \rho(\mathbf{x}) V(\mathbf{x}, t) - \mathbf{j}(\mathbf{x}) \mathbf{A}(\mathbf{x}, t) \right\} ,$$

where the $\mathbf{A}^2$ term of the non-relativistic theory was ignored, since it is of second order in the field. Prove the Kubo formula for the induced average current density

$$\langle j_\mu(\mathbf{x}, t) \rangle = \sum_{\nu=1}^{3} \int_{-\infty}^{t} dt' \int_{0}^{\beta} d\lambda \int d\mathbf{x}' \langle j_\nu(\mathbf{x}', -\imath \hbar \lambda) j_\mu(\mathbf{x}, t - t') \rangle_0 E_\nu(\mathbf{x}', t') .$$

© Springer International Publishing AG, part of Springer Nature 2018
L. A. Bányai, *A Compendium of Solid State Theory*,
https://doi.org/10.1007/978-3-319-78613-1_10

using the Kubo identity

$$\left[A, e^{-\beta H}\right] = e^{-\beta H} \int_0^\beta d\lambda e^{\lambda H} \left[H, A\right] e^{-\lambda H} .$$

(Check it by taking its matrix element between eigenstates of the Hamiltonian H.)
Consider the case of a homogeneous field in a homogeneous system introduced
adiabatically at $t = -\infty$ as e^{st} with $s \to +0$ to get the conductivity tensor

$$\sigma_{\mu\nu}(\omega) = \lim_{s \to +0} \lim_{\Omega \to \infty} \Omega \int_{-\infty}^0 dt e^{\imath\omega t} e^{st} \int_0^\beta d\lambda \langle j_\nu(t - \imath\hbar\lambda) j_\mu(0) \rangle_0 , \qquad (10.1)$$

where $j_\mu \equiv \frac{1}{\Omega} \int d\mathbf{x} j_\mu(\mathbf{x})$ and the infinite volume limit is to be understood in the
thermodynamic sense, i.e. at a fixed average carrier density. (It is understood, that
the Kubo formula is valid only for Coulomb non-interacting electrons i.e. the electric
field is the total one.)

10.2   Ideal Relaxation

Remake the above derivation by adding an ideal relaxation term

$$-\imath\hbar \frac{R - R_0}{\tau}$$

to the Liouville equation. However, now starting in equilibrium at time $t = 0$ and
measuring at $t = \infty$ without any adiabaticity. You shall see, that you get the same
formula, with the adiabatic parameter s replaced by $\frac{1}{\tau}$.
    Now, consider that the unperturbed Hamiltonian H is just the one describing the
motion in a constant magnetic field in the Landau gauge i.e. compute the matrix
elements with the Landau states and perform all the integrals. Do not forget, that the
velocity in the presence of the magnetic field is $\dot{\mathbf{x}} \equiv \frac{1}{m}(\mathbf{p} - \frac{e}{c}\mathbf{A}(\mathbf{x}))$!
    You shall obtain the same expression for the conductivity tensor as we derived in
Sect. 5.3 from the Boltzmann equation under an analogous ideal relaxation assumption.

10.3   Rate Equation for Bosons

Derive the transition rates for massive bosons interacting with acoustic phonons
using the many-body "golden rule". Show the detailed balance relation. Within the
approximation

$$\langle n_i n_j \rangle \approx \langle n_i \rangle \langle n_j \rangle$$

formulate the corresponding rate equation with discrete wave vectors $\mathbf{k}$. Analyze its properties. Perform the thermodynamic limit and remark the problem with the particle conservation if the concentration of bosons exceeds the critical one.

10.4 Bose Condensation in Time

Try to repair the above derived rate equation for massive bosons by implementing the idea of a macroscopic number of bosons in the $\mathbf{k} = 0$ state we discussed in the equilibrium theory of Bose condensation Sect. 7.2. Also consider an independent macroscopic degree of freedom $\langle n_0 \rangle$ proportional to the volume and separate from the discrete sums this contribution. Admit that only the rest converges to a Riemann integral in the thermodynamic limit. Analyze the properties of the new system of equations. Show, that it has an equilibrium attractor for $t \rightarrow \infty$ with or without condensate, depending on the total density. Remark the need for an arbitrary small condensate at $t = 0$ ("condensation kernel") to get a solution above the critical density.

10.5 Bose Condensation in a Finite Potential Well

The standard theory of Bose condensation implies the thermodynamic limit and describes an infinite homogeneous system. On the other hand, experimental evidence occurs in finite (confined) systems. One may try a quasi-classical approach to bosons in a finite (but not quantizing !) potential well to understand this aspect.

Consider the motion in the presence of a finite in depth and width, spherically symmetric, attractive ($v > 0$) potential well

$$U_0(r) = -v \left(1 - \left(\frac{r}{R} \right)^2 \right) \theta(R - r)$$

($r \equiv |\mathbf{x}|$) embedded in an infinite volume.

The corresponding equilibrium distribution is

$$f(\mathbf{p}, \mathbf{x}) = \frac{1}{e^{\beta(\frac{p^2}{2m} + U(\mathbf{x}) - \mu)} - 1} \; .$$

with the chemical potential determined by the total number of bosons

$$\int d\mathbf{x} \int \frac{d\mathbf{p}}{(2\pi \hbar)^3} f(\mathbf{p}, \mathbf{x}) = \langle N \rangle \; .$$

However, since one expects, that at low temperatures many bosons drop in the well, one has to take into consideration, that being close to each other they will interact. Usually this interaction is repulsive. Consider just a contact interaction potential $v(\mathbf{x}) = w\delta(\mathbf{x})$ ($w > 0$), that one may treat in a self-consistent manner by the effective potential

$$U(\mathbf{x}) = U_0(\mathbf{x}) + \int d\mathbf{x}' v(\mathbf{x} - \mathbf{x}') \int \frac{d\mathbf{p}}{(2\pi\hbar)^3} f(\mathbf{p}, \mathbf{x}') \ .$$

This potential is just a constant outside the well

$$U(\mathbf{x}) = w\langle n\rangle \qquad (for \quad r > R)$$

where $\langle n\rangle$ is the average density of bosons

$$\langle n\rangle = \int \frac{d\mathbf{p}}{(2\pi\hbar)^3} \frac{1}{e^{\beta(\frac{p^2}{2m}+w\bar{n}-\mu)} - 1} \ .$$

(A finite number of bosons in the well do not contribute to this equation.) A solution exists only for $\mu < wn_c$, where n_c is the critical density

$$n_c = \int \frac{d\mathbf{p}}{(2\pi\hbar)^3} \frac{1}{e^{\beta\frac{p^2}{2m}} - 1} \ .$$

Above this density an overall condensate should appear.

Now, it is most interesting to follow the scenario before this overall condensation occurs. For $r < R$ one gets a radius dependent density and the self-consistency equation

$$U(r) = U_0(r) + w \int \frac{d\mathbf{p}}{(2\pi\hbar)^3} \frac{1}{e^{\beta(\frac{p^2}{2m}+U(r)-\mu)} - 1} ; \qquad (r < R)$$

So long $U(r) - \mu > 0$, this equation has a solution, but afterwards, obviously not. Therefore, it is reasonable to correct this equation by admitting the possibility of a local condensate density $n_0(r)$ not included in the Bose distribution

$$U(r) = U_0(r) + w \left(n_0(r) + \int \frac{d\mathbf{p}}{(2\pi\hbar)^3} \frac{1}{e^{\beta(\frac{p^2}{2m}+U(r)-\mu)} - 1} \right) \qquad (for \quad r < R) \ .$$

The potential $U(r)$ varies monotonously and coming down from the top of the potential one might reach a radius $r_0 < R$, by which indeed

$$U(r_0) = \mu$$

and the integral over the Bose function reaches its maximal value. Due to the condition $U(r) - \mu \geq 0$ also for the points $r < r_0$, it must belong to the minimum of $U(r)$ and a condensate $n_0(r)$ must emerge

$$n_0(r) = \frac{1}{w} \left(U(r_0) - U_0(r) \right) \quad (for \quad 0 < r < r_0) .$$

Solve numerically the self-consistency equation for $r_0 < r < R$ (before the apparition of the condensate) to confirm this scenario. Since the transcendental self-consistency equation is local, for a numerical solution it is convenient to solve it in favor of U_0 at a given U, thus performing a simple integration over the momenta $\mathbf{p}$. The association to a certain radius r is given then by the explicit definition of $U_0(r)$.

Printed in the United States
By Bookmasters